Lecture Notes in Civil Engineering 775

Lecture Notes in Civil Engineering (LNCE) publishes the latest developments in Civil Engineering—quickly, informally and in top quality. Though original research reported in proceedings and post-proceedings represents the core of LNCE, edited volumes of exceptionally high quality and interest may also be considered for publication. Volumes published in LNCE embrace all aspects and subfields of, as well as new challenges in, Civil Engineering. Topics in the series include:

- Construction and Structural Mechanics
- Building Materials
- Concrete, Steel and Timber Structures
- Geotechnical Engineering
- Earthquake Engineering
- Coastal Engineering
- Ocean and Offshore Engineering; Ships and Floating Structures
- Hydraulics, Hydrology and Water Resources Engineering
- Environmental Engineering and Sustainability
- Structural Health and Monitoring
- Surveying and Geographical Information Systems
- Indoor Environments
- Transportation and Traffic
- Risk Analysis
- Safety and Security

To submit a proposal or request further information, please contact the appropriate Springer Editor:
- Pierpaolo Riva at pierpaolo.riva@springer.com (Europe and Americas);
- Swati Meherishi at swati.meherishi@springer.com (Asia—except China, Australia, and New Zealand);
- Wayne Hu at wayne.hu@springer.com (China).

All books in the series now indexed by Scopus and EI Compendex database!

Andrius Jurelionis · Paris A. Fokaides ·
Livio Mazzarella · Timo Hartmann
Editors

Building Digital Twins

Proceedings of BDTSC 2025

 Springer

Editors
Andrius Jurelionis
Faculty of Civil Engineering and Architecture
Kaunas University of Technology
Kaunas, Lithuania

Paris A. Fokaides
School of Engineering
Frederick University
Nicosia, Cyprus

Livio Mazzarella
Dipartimento di Energia
Politecnico di Milano
Milan, Italy

Timo Hartmann
Contecht GmbH
Berlin, Germany

ISSN 2366-2557   ISSN 2366-2565 (electronic)
Lecture Notes in Civil Engineering
ISBN 978-3-032-09039-3   ISBN 978-3-032-09040-9 (eBook)
https://doi.org/10.1007/978-3-032-09040-9

This Springer imprint is published by the registered company Springer Nature Switzerland AG
The registered company address is: Gewerbestrasse 11, 6330 Cham, Switzerland

If disposing of this product, please recycle the paper.

Preface

It is with great pleasure that I present the proceedings of the 1st International Conference on Building Digital Twin and Smart Cities (BDTSC), hosted at Kaunas University of Technology (KTU) in 2025. As Chairman of this inaugural edition, I am honoured to introduce this collection of peer-reviewed contributions that reflect the latest advancements and emerging research directions in the field of digital twins for the built environment.

The BDTSC conference was established to respond to the accelerating digital transformation of our buildings, cities, and infrastructure. As society grapples with the challenges of climate change, urbanisation, energy efficiency, and citizen wellbeing, digital twins are becoming key enablers of smarter, more sustainable, and data-driven decision-making. BDTSC was conceived as a multidisciplinary platform to convene researchers, practitioners, and policymakers around the shared objective of advancing digital twin methodologies, tools, and implementations for buildings and cities.

This conference was organised under the framework of the SmartWins project, a Horizon Europe Twinning initiative (Grant Agreement No. 101078997) coordinated by Kaunas University of Technology, which aims to enhance KTU's research excellence in the domain of digital twins for the built environment.

The contributions presented in this volume are the result of a highly competitive selection process and reflect the diversity of topics and approaches discussed during the event. A total of 15 papers are included, representing academic institutions, applied research centres, and innovation ecosystems from across Europe. Together, they offer a snapshot of current thinking and a springboard for further collaboration and exploration.

Several papers focus on frameworks and methodological innovations. The DWELT framework (Define, Wire, Engineer, Leverage, Transfer) proposes a structured path towards scalable and interoperable digital twin solutions through a System-of-Systems lens. Another study evaluates the compatibility of TEASER and AixLib libraries in the Modelon Impact environment, offering insights into functional interoperability of physics-based models. Meanwhile, a paper exploring IFC-based open BIM models on web platforms demonstrates the importance of openness and long-term data reliability for sustainable digital workflows.

Urban-scale applications of digital twins were also strongly represented. Two papers provide complementary perspectives: one defines data requirements and a reference architecture for green spaces and ecosystems in urban digital twins (UDTs), while the other outlines a development agenda for UDTs, focusing on modularity, presentation-agnosticism, and stakeholder inclusion.

Artificial intelligence (AI) as a cross-cutting enabler is explored in an overview paper mapping the use of AI tools across the building lifecycle—from design and planning to operation, optimisation, and compliance. The role of co-creation tools in sustainable building design is also highlighted, with participatory approaches positioned as a means to enhance environmental performance and social value.

Several papers offer tangible case studies. A study from the HYCOOL-IT project presents a digital twin for a data centre integrating waste heat recovery and predictive control strategies. Another case from the HPC4AI data centre in Turin documents the implementation of experimental server cooling systems and management platforms, reinforcing the research-to-application link.

Innovation in 3D data acquisition is also featured. One contribution introduces a dynamic method for automated planning and re-planning of terrestrial laser scanning (TLS) using discretised key-points and next-best-view algorithms. In parallel, a study focusing on a building's heat storage system illustrates how numerical models integrated into a digital twin can optimise heating performance and reduce energy loss.

Digitalisation in heritage conservation is addressed through a paper on sustainable renovation of a cultural heritage building in Kaunas, combining HBIM, environmental monitoring, and low-carbon material selection. This intersection of technology, history, and sustainability reflects an increasingly important research direction.

Training and user engagement are also critical components of the digital twin ecosystem. A study on XR technologies in high-risk industrial training offers a methodological framework for evaluating training effectiveness and instructor adoption. In a different domain, the DCU Campus Explorer illustrates how a digital twin can enhance citizen engagement and inclusion across a university campus, supporting access to wellness services, quiet spaces, and real-time navigation.

Finally, a study on a knowledge-based configuration expert system offers a solution for empowering non-expert users – such as apartment managers – in early-stage renovation planning. The proposed system, tested in the Estonian context, bridges the gap between technical complexity and user-friendly decision support.

On behalf of the organising committee, I would like to extend my sincere thanks to all authors, reviewers, session chairs, and participants who contributed to the success of the conference. Special recognition is due to Kaunas University of Technology for its vision and support in hosting this first edition and to the SmartWins consortium for fostering the collaborative environment that made this gathering possible.

It is my hope that this volume will serve as a valuable reference and inspiration for future research and innovation in digital twins and smart cities.

Paris A. Fokaides

Organization

Committees

Paris A. Fokaides	Kaunas University of Technology, Lithuania; Frederick University, Cyprus
Andrius Jurelionis	Kaunas University of Technology, Lithuania
Livio Mazzarella	Politecnico di Milano, Italy
Timo Hartmann	Technical University of Berlin, Germany

Scientific Committee

Paris A. Fokaides (Chair of Committee)	Kaunas University of Technology, Lithuania; Frederick University, Cyprus
Ali Intizar	Dublin City University, Ireland
Andrius Jurelionis	Kaunas University of Technology, Lithuania
Angelo Ciribini	Università degli Studi di Brescia, Italy
Darius Pupeikis	Kaunas University of Technology, Lithuania
Dimosthenis Ioannidis	Centre for Research and Technology-Hellas, Greece
Edmond Saliklis	California Polytechnic State University, USA
Farzad Pour Rahimian	Teesside University, UK
Ioannis Brilakis	University of Cambridge, UK
Kjeld Svidt	Aalborg University, Denmark
Lavinia Chiara Tagliabue	University of Turin, Italy
Livio Mazzarella	Politecnico di Milano, Italy
Qian Wang	KTH Royal Institute of Technology, Sweden
Rossano Scoccia	Politecnico di Milano, Italy
Sanju Tiwari	Sharda University, Greater Noida, India
Targo Kalamees	Tallinn University, Estonia
Timo Hartmann	Technical University of Berlin, Germany
Vangelis Angelakis	Linköping University, Sweden
Vishal Singh	Indian Institute of Science, India
Pieter Pauwels	Eindhoven University of Technology, Netherlands
Pedro Meda Magalhães	University of Porto, Portugal

Organizing Committee

Lina Morkūnaitė	Kaunas University of Technology, Lithuania
Gintarė Stankevičiūtė	Kaunas University of Technology, Lithuania
Paulius Spūdys	Kaunas University of Technology, Lithuania
Iryna Osadcha	Kaunas University of Technology, Lithuania
Viktoras Jasaitis	Kaunas University of Technology, Lithuania
Aušra Andriukaitienė	Kaunas University of Technology, Lithuania
Aušra Mlinkauskienė	Kaunas University of Technology, Lithuania

Building Digital Twin Scientific Conference (BDTSC) 2025 (https://bdtsc.ktu.edu/bdtsc-2025/)

Contents

Design Wire Engineer Leverage and Transfer (DWELT) Framework for Building-Level Digital Twins

Karim Farghaly[1](✉) , Pedro Mêda[2] , Conor Shaw[3] , James O'Donnell[3] ,
Fulvio Re Cecconi[4] , and Nicola Moretti[1]

[1] Bartlett School of Sustainable Construction, University College London,
1-19 Torrington Place, London WC1E 7HB, UK
`Karim.farghaly@ucl.ac.uk`
[2] ICS/CONSTRUCT/GEQUALTEC, Faculty of Engineering, University of Porto, 4200-465
Porto, Portugal
[3] School of Mechanical and Materials Engineering, University College Dublin,
Engineering Building Belfield Dublin 4, Dublin, Ireland
[4] Department of Architecture Built Environment and Construction Engineering, Politecnico Di
Milano, Via Ponzio 31, 20133 Milano, Italy

Abstract. The adoption of Digital Twins (DTw) related technologies in the Architecture, Engineering, Construction, and Operations (AECO) sector has enhanced data-driven decision-making through bi-directional data flows and real-time analytics. However, the risk of creating new data silos and interoperability hurdles with other technologies and initiatives persist. To unlock the full potential of DTw, integration with broader digital initiatives – such as Digital Building Logbooks and Digital Product Passports – is essential. Addressing these challenges requires a System-of-Systems (SoS) approach and coordinated processes. This research explores the transition towards a SoS approach in the built environment sector. A qualitative methodology was employed, using co-creation workshop with industry professionals, software providers, and academic experts from diverse regions. The data gathered were analyzed using grounded theory, leading to the development of a scalable framework for DTw solutions. This study introduces the Define-Wire-Engineer-Leverage-Transfer (DWELT) framework, designed to support organizations in effectively implementing DTw through a SoS approach.

Keywords: Digital Twins · System of Systems · Interoperability · Ontology · BIM · Plug and Play

1 Introduction

The built environment sector is responsible for approximately 36% of global energy consumption and nearly 39% of CO_2 emissions [1]. In Europe, increasing regulatory pressure through frameworks – such as the Energy Performance of Buildings Directive (EPBD), the New Construction Product Regulations, the Renovation Wave, Ecodesign

© The Author(s) 2026
A. Jurelionis et al. (Eds.): BDTIC 2025, LNCE 775, pp. 1–12, 2026.
https://doi.org/10.1007/978-3-032-09040-9_1

for Sustainable Products Regulation (ESPR) and the EU Taxonomy for Sustainable Activities – aims to drive sustainability, energy efficiency, and circularity across the building life cycle. Digital advancements in Building Information Modelling (BIM) first and more recently in Digital Twins (DTw) have significantly improved data management, real-time monitoring, and predictive analytics, supporting these sustainability goals [2, 3]. However, current DTw implementations in the AECO industry are predominantly centralized, leading to issues such as data fragmentation, poor interoperability, and limited accessibility, especially for Small and Medium-sized Enterprises [4, 5]. The reliance on monolithic digital solutions restricts scalability and hinders the ability to capture the full life-cycle benefits of DTw. Furthermore, existing approaches often fail to support the dynamic and interconnected nature of modern construction environments, where multiple stakeholders, platforms, and systems must collaborate effectively [6]. Consequently, despite their potential, current implementations risk creating new silos between data, technologies, and initiatives. Overcoming these barriers requires a shift from isolated DTw implementations to a more decentralized, System-of-Systems (SoS) approach that enables cross-platform interoperability, open data exchange, and integration with regulatory compliance mechanisms. To maximize the impact of digitalization in the built environment, DTw must integrate seamlessly with other digital frameworks, such as Digital Building Logbooks (DBLs) and Digital Product Passports [7]. Yet, achieving this level of integration remains a major challenge.

This research aims to address these challenges through the development of the Define-Wire-Engineer-Leverage-Transfer (DWELT) framework, which leverages a SoS approach to create a scalable, interoperable, and decentralized DTw ecosystem. By moving beyond traditional centralized models, DWELT enables more effective DTw adoption, enhances accessibility for SMEs, and facilitates compliance with regulations. The framework provides a strategic pathway to achieving a fully integrated digital built environment, ultimately accelerating the transition to a sustainable, energy-efficient, and circular sector.

2   Literature Review

2.1   Digital Twins for Buildings

The AECO industry is undergoing a significant transformation driven by the adoption of advanced digital technologies, particularly BIM and DTw [3, 8]. BIM has played a crucial role in enhancing efficiency, accuracy, and collaboration by providing a structured digital representation of buildings throughout their lifecycle. However, the emergence of DTw marks the next phase in this digital evolution, offering real-time, data-driven replicas of physical assets that enable predictive maintenance, real-time monitoring, and improved decision-making [9, 10]. A DTw can be defined in extreme summary as a spatial-temporal virtual representation of a physical entity or system that integrates all relevant static and dynamic information throughout its lifecycle. It connects the physical and virtual entities via data connections, enabling capabilities such as monitoring, analysis, decision-making, and predictive modeling [11, 12].

A core aspect of DTw in the AECO sector is their ability to facilitate bidirectional data exchange between physical and virtual environments, progressing through distinct stages: Digital Models (no data exchange), Digital Shadows (unidirectional data exchange), and fully integrated Digital Twins (bidirectional data exchange) [13]. The effectiveness of DTw depends on their integration with evolving technologies, such as sensor networks and semantic information processing [14, 15]. However, implementing DTw at scale is hindered by the complexity of construction projects, the unpredictability of implementation cost [16] and clear human-data interaction (HDI) [17], and the lack of standardized frameworks for interoperability across different systems [18]. While a Common Data Environment (CDE) can help consolidate information from various sources, the coexistence of multiple digital concepts – such as BIM, Cyber-Physical Systems (CPS), Smart Buildings, Facilities Management Systems, and DTw – creates ambiguity in deployment strategies [19].

The industry's struggle to define a clear and scalable approach to twinning has led to fragmented implementations, limiting the potential of DTw [6]. This can be attributed to the fact that current implementations remain largely centralized, restricting scalability, flexibility, and system-wide interoperability. Many challenges – such as data silos and platform incompatibility – stem from the industry's reliance on centralized solutions that fail to support the dynamic and interconnected nature of modern AECO environments [20]. A shift towards decentralized, system-of-systems (SoS) approaches is necessary to unlock the full potential of DTw, allowing multiple platforms and stakeholders to seamlessly integrate and collaborate.

2.2 System of Systems

System-of-Systems Engineering (SoSE) emerged as a research discipline from general systems theory and systems engineering to address the increasing complexity of interconnected systems. General systems theory, introduced by [21], emphasizes a holistic approach to studying relationships and interactions within complex systems. The most critical factors in SoSE identified by the literature are the integration of multiple independent systems, understanding their interactions and interdependencies, achieving a unified goal beyond individual capabilities, managing system complexity and emergent behaviors, and recognizing the operational and managerial independence of constituent systems [22]. Over the past two decades, SoSE has expanded beyond military domain, becoming a fundamental methodology in various sectors. The widespread adoption of SoSE highlights its increasing relevance in managing complex, multi-domain, and continuously evolving systems.

A System of Systems (SoS) refers to an ecosystem of interconnected, collaborative, and interactive subsystems that retain operational independence while contributing to a larger function. Unlike traditional hierarchical systems, a SoS is characterized by its openness at multiple levels. It remains open at the top, meaning there is no predefined overarching application, allowing new applications to emerge dynamically. It is also open at the bottom, where system components are defined functionally rather than concretely, ensuring flexibility in integration [23]. From an engineering perspective, a fundamental question is how SoS can be designed to achieve desired emergent properties [24]. To better understand the interactions between SoS components, [25] proposed a generic representation of SoS that distinguishes between central Systems of Interest (SoI) and Wider Systems of Interest (WSoI). The central SoI consists of core subsystems essential for achieving the desired emergent behavior, forming the foundation of the SoS. In contrast, the WSoI comprises systems that exist within the SoS environment and can enhance emergent properties when integrated but are not essential to core functionality [24]. This distinction helps clarify how different components within a SoS interact and contribute to overall system performance.

2.3 Synthesis

This inherent openness enables an SoS to support innovation, adaptability, and scalability, making it particularly suitable for complex environments such as DTw, smart cities, and integrated data ecosystems in the AECO sector. However, additional research is needed to understand the relationships between DTw, associated technologies, systems, and information containers (i.e., the SoI), as well as the WSoI, such as digital building logbooks and digital product passports. Therefore, this paper aims to develop a framework to address this knowledge gap and proposes system design and architecture to achieve scalable, interoperable DTw.

3   Methodology

This research employs a co-creation approach to develop a robust framework that facilitates the integration of building information systems through a SoSE perspective. Co-creation is an innovative-driven research approach where problem-owners and end-users actively participate in designing and developing solutions, rather than merely identifying problems. By involving key stakeholders, this methodology ensures that the proposed framework is practical, scalable, and aligned with industry needs. This study follows an engaged scholarship approach [26], where the research team is embedded in real-world discussions and decision-making processes. The co-creation process was conducted through a workshop and six structured meetings, engaging software providers, construction industry experts, and academic researchers from diverse backgrounds. These interactions facilitated the collaborative development of a framework that is both theoretically sound and practically applicable. Table 1 provides an overview of the stakeholders participating in the workshop activities, including their affiliations and years of industry experience, ensuring a balanced representation of expertise.

To systematically analyze the data gathered during the co-creation sessions, the study adopted a Grounded Theory Approach [27]. This qualitative research method is well-suited for generating theory from systematically collected and analyzed data, making it ideal for exploring the challenges, opportunities, and strategies associated with BIM and DTw integration in construction. The grounded theory methodology involved continuous data collection and analysis, where data was coded and categorized into emerging themes. This iterative process allowed the framework to be refined based on real-world insights and industry feedback, ensuring its adaptability to various construction project scenarios. By integrating co-creation with a Grounded Theory Approach, this research not only develops a scalable and implementable framework but also contributes to the broader discourse on DTw adoption in the built environment. The methodology ensures that the framework is not just an academic construct but a practical tool that can drive innovation, improve decision-making, and enhance interoperability in construction projects.

Table 1. Co-creation workshop Participants

Participant No.	Organization Type	Job Title	Years of Experience
1	University 1	Associate Professor	10–20
2	University 1	Assistant Professor	5–10
3	University 1	Reader	10–20
4	Consultancy	Data Consultant	>20
5	Software Provider 1	CTO	>20
6	Software Provider 1	CEO	>20
7	Software Provider 2	CTO	>20
8	Software Provider 2	CEO	>20
9	Software Provider 3	CTO	10–20
10	Software Provider 4	CEO	>20
11	University 2	Associate Professor	10–20
12	University 3	Associate Professor	10–20
13	University 3	PhD student	<5
14	Real Estate Provider	Innovation Manager	10–20
15	University 4	Professor	10–20

4   DWELT System Architecture

The DWELT system architecture provides a structured, multi-layered framework for developing scalable and interoperable DTw in the AECO industry. Addressing key challenges such as data fragmentation, interoperability issues, and integration with existing platforms, the architecture is designed to enable seamless information exchange and advanced analytics while ensuring adaptability across diverse use cases. By incorporating a federated data model, validation mechanisms, and user-centric interaction layers, DWELT offers a comprehensive approach to building and managing DTw solutions. The framework is structured into five interconnected layers (Fig. 1), each serving a critical function in the DTw development process:

The **Data Requirements and Specifications Layer** forms the foundation of the DWELT system architecture by ensuring that data structures are clearly defined, interoperable, and aligned with industry standards. This layer focuses on requirement identification, semantic enrichment and semantic alignment, establishing a structured approach to integrating diverse data sources into a federate DTw ecosystem. A key aspect of this layer is requirement identification, which follows a combined top-down and bottom-up approach [28]. The top-down process ensures that high-level regulatory, environmental, and business objectives – such as compliance with regulations and sustainability standards – are embedded within the data framework. Simultaneously, the bottom-up approach gathers detailed technical and operational needs directly from stakeholders, including building owners, operators, contractors, and service providers. This dual perspective ensures that the data framework is both strategically aligned and practically implementable, enabling scalable and adaptable DTw solutions. To meet these diverse requirements, semantic enrichment is applied, enhancing data interoperability by structuring and standardizing information exchange. By leveraging linked data principles and metadata schemas, this process ensures that data carries contextual meaning, allowing systems to interpret and integrate it effectively. The final step in this layer is semantic alignment, which harmonizes federated data models with a unified ontology [29], ensuring data consistency, integration, and comparability across different platforms and demonstrator sites. This layer of the DWELT architecture, provides the necessary foundation for seamless data interoperability, automated validation, and cross-platform integration, enabling a robust and scalable DTw ecosystem.

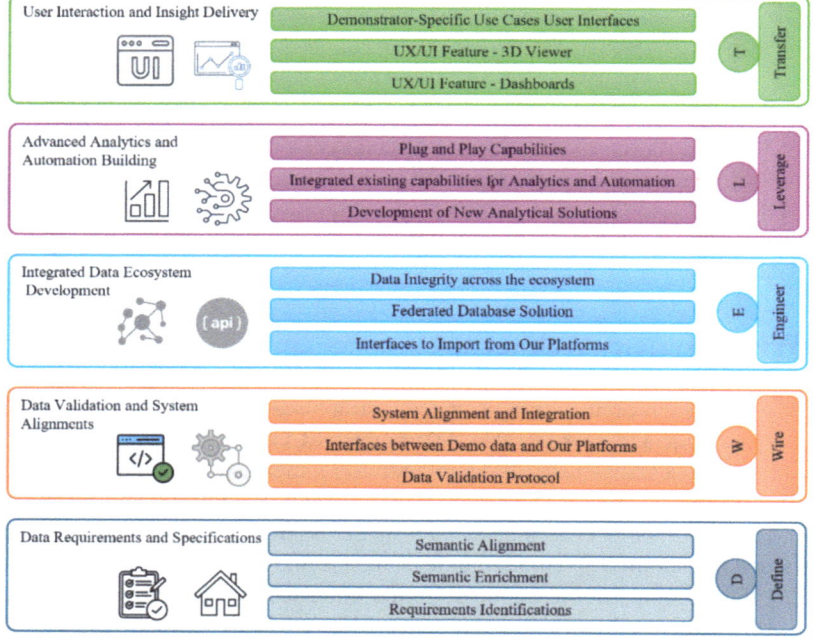

Fig. 1. The DWELT system architecture framework

The **Data Validation and System Alignment Layer** establishes mechanisms for validating, structuring, and integrating heterogeneous data sources, enabling seamless interoperability between BIM platforms, IoT sensor networks, energy monitoring systems, and lifecycle assessment tools. Given the complexity of data inputs – ranging from structured BIM/Industry Foundation Classes (IFC) models to unstructured PDFs, spreadsheets, and real-time sensor data [30] – this layer ensures that all information meets the necessary quality standards before being ingested into the DWELT ecosystem. To achieve system-wide alignment, this layer implements a top-down validation framework that applies predefined quality control rules – such as data type verification, cross-field consistency, and anomaly detection – while simultaneously leveraging a bottom-up approach to refine validation criteria based on real-world data sources and stakeholder needs.

Data validation protocols employ techniques such as Information Delivery Specifications (IDS) and Segment Protocols, ensuring that incoming data is structured, standardized, machine-readable, and ready for integration into analytical and automation processes. Beyond validation, system alignment and data integration mechanisms facilitate bidirectional data exchange between DTw and existing software platforms. This is achieved through Application Programming Interfaces (APIs) for direct system connections, Publish/Subscribe (Pub/Sub) messaging patterns for real-time updates, and semi-automated data ingestion workflows that balance automation with human oversight for data-intensive processes. By enforcing data validation, enabling efficient system integration, and ensuring real-time interoperability, this layer eliminates data silos, enhances cross-platform functionality, and establishes a scalable foundation for advanced analytics and automation in DTw applications.

The **Integrated Data Ecosystem Layer** forms the core infrastructure of the DWELT architecture, ensuring data integrity, scalability, and interoperability across diverse digital solutions. A fundamental challenge in DTw ecosystems is ensuring that data remains accurate, transparent, and consistent across different systems. This layer incorporates ledger-based verification mechanisms, creating an immutable record for data transactions to enhance trust and accountability. By aligning with FAIR (Findability, Accessibility, Interoperability, and Reusability) principles and European data governance strategies [31], the system ensures compliance with industry standards while providing a robust foundation for secure and structured data management. At the heart of this layer is the design and implementation of a federated database, which unifies diverse data models through a graph database. This structure enables advanced query capabilities, semantic search functionalities, and efficient discovery of relationships between data points, supporting interoperability across various DTw applications. Furthermore, customizable data import interfaces allow for flexible data ingestion from multiple sources – including sensor feeds, BIM platforms, and structured documents – ensuring that stakeholders can seamlessly integrate and access critical project information. By creating a modular, scalable, and secure data ecosystem, this layer eliminates data silos, enhances cross-platform compatibility, and establishes a reliable foundation for analytical and automation processes. Its federated approach ensures that DTw solutions remain adaptable, resilient, and accessible, fostering greater collaboration and innovation across the AECO industry.

The **Advanced Analytics and Automation Building Layer** enhances the capabilities of DTw by integrating existing analytical frameworks with new Artificial Intelligence-driven (AI-driven) automation solutions in a plug-and-play architecture. This approach ensures that existing off-the-shelf digital tools, platforms, and services can be seamlessly integrated, while newly developed analytics solutions align with the requirements defined in the Data Requirements and Specifications Layer (Define – D). By structuring the system in this way, DWELT enables flexibility, scalability, and adaptability, ensuring that stakeholders can leverage both pre-existing industry solutions and emerging technologies to enhance decision-making processes. A key aspect of this layer is the development of new analytical solutions based on the structured, semantically enriched data environment facilitated by the DWELT architecture. Unlike traditional DT implementations, which often suffer from low-quality, unstructured, and siloed data, DWELT's federated ecosystem allows for high-quality, standardized data exchange, ensuring predictive maintenance, energy consumption forecasting, lifecycle tracking, and algorithmic control of building systems. Machine Learning (ML) models, synthetic data generation, and scenario analysis tools are developed within this structured ecosystem, enabling data-driven automation strategies that adapt to building-specific conditions and stakeholder-defined needs.

Beyond developing new tools, this layer also emphasizes leveraging and integrating existing capabilities to meet automation and intelligence requirements in a modular, plug-and-play fashion. Key functionalities include a query engine for real-time insights, access control mechanisms, platform APIs for data integration, and predictive feedback loops that continuously refine automation processes. These elements work together to create a scalable and standards-aligned analytics platform, enhancing sustainability, operational efficiency, and transparency across DT applications. By embedding plug-and-play analytics, AI-driven automation, and a modular intelligence layer, DWELT ensures that stakeholders can easily integrate, customize, and scale their DT solutions while maintaining interoperability with pre-existing and newly developed tools. This architecture bridges the gap between raw data and actionable insights, providing a robust foundation for energy efficiency, circularity, and optimized building performance in a rapidly evolving AECO industry.

The **User Interaction and Insight Delivery Layer** ensures that DT solutions are accessible, intuitive, and user-driven, bridging the gap between technical insights and end-user needs. This layer enables interactive exploration, visualization, and decision-making through customizable dashboards, 3D viewers, and demonstrator-specific interfaces. A key component of this layer is the development of dynamic dashboards, designed to provide real-time monitoring, performance tracking, and actionable insights. These dashboards incorporate high interactivity, filtering, and data exploration functionalities, allowing users to engage with information in a meaningful way. The integration of decision-support tools further enables building performance analysis, predictive maintenance tracking, and sustainability reporting, ensuring that stakeholders can make informed, data-driven decisions. Complementing the dashboards, the 3D visualization interface enhances user engagement by offering real-time, web-based rendering of building models and geospatial data without requiring additional software plugins. By leveraging Web3D, PostgreSQL for data management, and Three.js for rendering, the 3D viewer supports seamless interaction between users and complex DT environments, making it easier to navigate, assess, and optimize building performance. The data exchange between the front-end and back-end follows standardized formats, ensuring interoperability with other DWELT components. The final aspect of this layer involves the customization and deployment of use case specific interfaces, ensuring that each user environment is tailored to the specific needs of building owners, operators, and stakeholders [17]. This ensures that the interfaces remain user-centric, flexible, and scalable while integrating both pre-existing solutions and newly developed automation capabilities from the Advanced Analytics and Automation Building Layer (Leverage – L). To ensure usability and adoption, a continuous validation and refinement process is embedded in this layer. Stakeholder engagement through focus groups, feedback loops, and qualitative research allows for iterative improvements, aligning the UX/UI components with project objectives and user requirements.

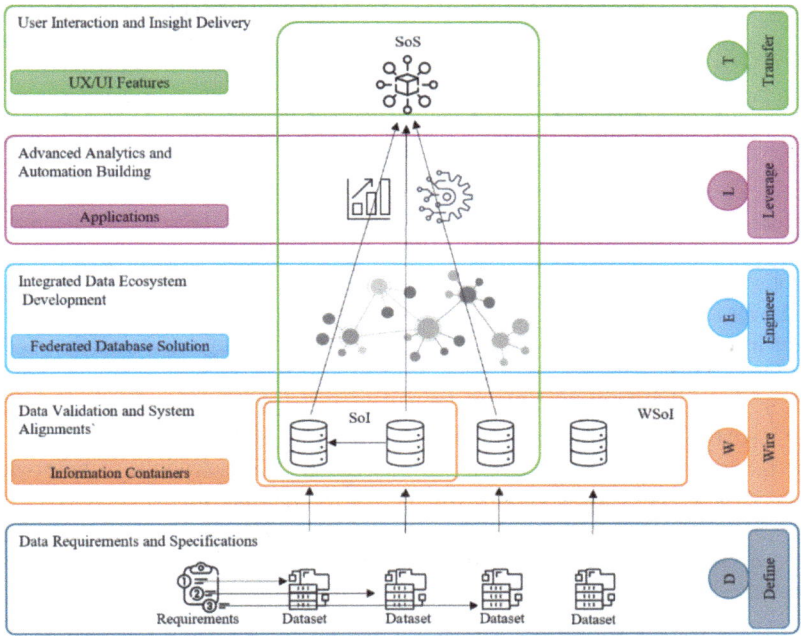

Fig. 2. The DWELT framework aligned with the SoSE.

The DWELT architecture demonstrates strong alignment with SoSE principles (Fig. 2), providing a foundation for scalable and interoperable SoS implementation within DTw across the AECO sector. The Data Validation and System Alignment Layer plays a pivotal role by incorporating both systems of interest (SoI)—such as BIM, IoT, and LCA tools—and wider systems of interest (WSoI), including DBL, DPP, and regulatory compliance platforms. This inclusive scope supports the emergence of higher-order system behaviors while maintaining the operational independence of each subsystem. A critical prerequisite at this stage is the establishment of a clear, shared ontology through semantic enrichment and alignment. Once this is in place, the Advanced Analytics and Automation Building Layer and the User Interaction and Insight Delivery Layer can fully realize the system-of-systems concept, enabling plug-and-play capabilities, adaptive intelligence, and user-centric interaction. These layers support openness at the top—allowing new applications to emerge – and at the bottom – enabling flexible integration of diverse functional components – thereby fulfilling SoSE criteria for openness, emergent behavior, and dynamic scalability within complex DTw ecosystems.

5   Conclusions

The DWELT framework presents a scalable, interoperable, and decentralized approach to DTw implementation in the built environment, addressing data fragmentation, poor interoperability, and limited accessibility. By adopting SoS architecture, DWELT enables seamless integration between existing digital tools and new analytical solutions, ensuring that DT ecosystems are modular, adaptable, and aligned with evolving industry requirements. The five-layered architecture forms the foundation of this approach, beginning with the Data Requirements and Specifications Layer (Define – D), which establishes a federated data model through a top-down and bottom-up approach to requirement identification. The Data Validation and System Alignment Layer (Wire – W) ensures that data is structured, validated, and interoperable, enabling seamless integration between heterogeneous sources. The Integrated Data Ecosystem Layer (Engineer – E) provides a secure and federated database solution, facilitating real-time data exchange through APIs and publish/subscribe mechanisms. The Advanced Analytics and Automation Building Layer (Leverage – L) enhances decision-making and operational efficiency by leveraging machine learning, scenario analysis, and predictive modeling, ensuring that DT solutions evolve with data-driven automation. Finally, the User Interaction and Insight Delivery Layer (Transfer – T) ensures stakeholder engagement through intuitive dashboards and 3D visualizations, bridging technical insights with end-user needs. By embedding plug-and-play approach and semantic enrichment, DWELT provides a comprehensive framework for implementing DTw across the entire building lifecycle. Whilst preliminary comparisons with existing frameworks are discussed, a more extensive comparative analysis remains a limitation and will be addressed in future research directions. Compared to other architectures, the embedded continuous co-creation cycle ensures that the system remains aligned with industry regulations, sustainability goals, and real-world requirements, fostering widespread adoption and long-term impact. In addition to this, as the AECO industry continues its digital transformation, DWELT serves as a replicable and adaptable model, offering a future-proof foundation for AECO digital transformation. A key limitation is that the AECO sector is largely controlled by dominant software providers, which can hinder the adoption of open, interoperable frameworks like DWELT. Its full potential relies on broad stakeholder engagement and adherence to open-data principles. To demonstrate its value and practical applicability, the architecture will be validated through case studies.

References

1. The European Green Deal – European Commission. https://commission.europa.eu/strategy-and-policy/priorities-2019-2024/european-green-deal_en. Last accessed 29 Mar 2025
2. Deng, M., Menassa, C.C., Kamat, V.R.: From BIM to digital twins: A systematic review of the evolution of intelligent building representations in the AEC-FM industry. J. Inform. Technol. Construct. **26**, 58–83 (2021). https://doi.org/10.36680/J.ITCON.2021.005

3. Sacks, R., Brilakis, I., Pikas, E., Xie, H.S., Girolami, M.: Construction with digital twin information systems. Data-Centric Eng. (2020). https://doi.org/10.1017/dce.2020.16
4. Doe, R., Kaur, K., Selway, M., Stumptner, M.: Interoperability in AECO and the oil & gas sectors: object-based standards and systems. ITcon. **27**, 312–334 (2022). https://doi.org/10.36680/j.itcon.2022.016
5. Jahangir, M.F., Schultz, C.P.L., Kamari, A.: A review of drivers and barriers of Digital Twin adoption in building project development processes. ITcon. **29**, 141–178 (2024). https://doi.org/10.36680/j.itcon.2024.008
6. Tuhaise, V.V., Tah, J.H.M., Abanda, F.H.: Technologies for digital twin applications in construction. Autom. Constr. **152**, 104931 (2023). https://doi.org/10.1016/j.autcon.2023.104931
7. Mêda Magalhães, P., Calvetti, D., Kifokeris, D., Kassem, M.: A process-based framework for digital building logbooks. Presented at the EC3 Conference 2022 (2022). https://doi.org/10.35490/EC3.2022.183
8. Wang, K., Guo, F., Zhang, C., Schaefer, D.: From Industry 4.0 to Construction 4.0: barriers to the digital transformation of engineering and construction sectors. Eng. Construct. Architect. Manage. **31**, 136–158 (2024). https://doi.org/10.1108/ECAM-05-2022-0383
9. AlBalkhy, W., Karmaoui, D., Ducoulombier, L., Lafhaj, Z., Linner, T.: Digital twins in the built environment: Definition, applications, and challenges. Autom. Construct. (2024). https://doi.org/10.1016/j.autcon.2024.105368
10. ISO/IEC 30173:2023. https://www.iso.org/standard/81442.html. Last accessed 29 Mar 2025
11. Abdelrahman, M., Macatulad, E., Lei, B., Quintana, M., Miller, C., Biljecki, F.: What is a Digital Twin anyway? Deriving the definition for the built environment from over 15,000 scientific publications. Build. Environ. **274**, 112748 (2025). https://doi.org/10.1016/j.buildenv.2025.112748
12. SPHERE: Digital Twin Definitions for Buildings. https://sphere-project.eu/download/sphere-digital-twin-definitions-for-buildings/. Last accessed 29 Mar 2025
13. Tchana, Y., Ducellier, G., Remy, S.: Designing a unique Digital Twin for linear infrastructures lifecycle management. Procedia CIRP. **84**, 545–549 (2019). https://doi.org/10.1016/j.procir.2019.04.176
14. Calvetti, D., Mêda, P., Hjelseth, E., de Sousa, H.: Incremental digital twin framework: a design science research approach for practical deployment. Autom. Constr. **170**, 105954 (2025). https://doi.org/10.1016/j.autcon.2024.105954
15. Mêda, P., Calvetti, D., Hjelseth, E., Sousa, H.: Incremental Digital Twin Conceptualisations Targeting Data-Driven Circular Construction. BUILDINGS. 11, (2021). https://doi.org/10.3390/buildings11110554
16. Bortolin, G., Moretti, N., Farghaly, K., Chen, W.: An Approach to the Estimation of Digital Twins Technological Components. Presented at the EC3 Conference 2024 (2024). https://doi.org/10.35490/EC3.2024.303
17. Soman, R.K., Farghaly, K., Mills, G., Whyte, J.: Digital twin construction with a focus on human twin interfaces. Autom. Constr. **170**, 105924 (2025). https://doi.org/10.1016/j.autcon.2024.105924
18. Moretti, N., Xie, X., Merino Garcia, J., Chang, J., Kumar Parlikad, A.: Federated data modeling for built environment digital twins. J. Comput. Civ. Eng. **37**, 04023013 (2023). https://doi.org/10.1061/JCCEE5.CPENG-4859
19. Akanmu, A.A., Anumba, C.J., Ogunseiju, O.O.: Towards next generation cyber-physical systems and digital twins for construction. Journal of Information Technology in Construction. 26, 505–525 (2021). https://doi.org/10.36680/j.itcon.2021.027
20. ISO/IEC TR 30172:2023 – Internet of things (IoT) — Digital twin — Use cases. https://www.iso.org/standard/81578.html. Last accessed 29 Mar 2025

21. Bertalanffy, L. von: General systems theory as integrating factor in contemporary science. Akten des XIV. Internationalen Kongresses für Philosophie. 2, 335–340 (1968)

22. Dahmann, J.: Current landscape of system of systems engineering. In: 2024 19th Annual System of Systems Engineering Conference (SoSE), pp. 1–9 (2024). https://doi.org/10.1109/SOSE62659.2024.10620929

23. Sadeghi, M., Carenini, A., Corcho, O., Rossi, M., Santoro, R., Vogelsang, A.: Interoperability of heterogeneous Systems of Systems: from requirements to a reference architecture. J. Supercomput. **80**, 8954–8987 (2024). https://doi.org/10.1007/s11227-023-05774-3

24. Dormeier, C., Mindt, N., Niemeyer, J.F., Asghari, R., Mennenga, M.: Review and framework for the engineering of Business Models for Sustainability: a System of Systems perspective. Sustain. Product. Consumpt. **51**, 1–22 (2024). https://doi.org/10.1016/j.spc.2024.08.030

25. Mennenga, M., Cerdas, F., Thiede, S., Herrmann, C.: Exploring the opportunities of system of systems engineering to complement sustainable manufacturing and life cycle engineering. Procedia CIRP **80**, 637–642 (2019). https://doi.org/10.1016/j.procir.2019.01.026

26. Voordijk, H., Adriaanse, A.: Engaged scholarship in construction management research: the adoption of information and communications technology in construction projects. Constr. Manag. Econ. **34**, 536–551 (2016). https://doi.org/10.1080/01446193.2016.1139145

27. Matavire, R., Brown, I.: Profiling grounded theory approaches in information systems research. Eur. J. Inf. Syst. **22**, 119–129 (2013). https://doi.org/10.1057/ejis.2011.35

28. Farghaly, K., Jones, K.: Enhancing requirement-information mapping for sustainable buildings: introducing the SFIR ontology. In: Advances in Conceptual Modeling, pp. 242–248. Springer Nature Switzerland, Cham (2023). https://doi.org/10.1007/978-3-031-47112-4_23

29. Farghaly, K., Soman, R., Whyte, J.: CSite ontology for production control of construction sites. Autom. Constr. **158**, 105224 (2024). https://doi.org/10.1016/j.autcon.2023.105224

30. ISO 16739–1:2024, https://www.iso.org/standard/84123.html, last accessed 2025/03/30

31. Communication from the Commission to the European Parliament, the Council, the European Economic and Social Committee and the Committee of the Regions A European Strategy For Data (2020)

Evaluating the Interoperability of TEASER and AixLib for Building Digital Twins Within Modelon Impact Environment: A Case Study

Laura Zabala Urrutia[1,2]($\boxtimes$), Sergiu Crisan[1], Estíbaliz Pérez Iribarren[1],
Iker González Pino[1], and Jesús Febres Pascual[2]

[1] Energy Engineering Department, Bilbao School of Engineering, University of the Basque Country UPV/EHU, Bilbao, Spain
laura.zabala@r2msolution.es
[2] R2M Solution Spain SL, Madrid, Spain

Abstract. The building sector's significant contribution to global energy consumption and greenhouse gas emissions necessitates more intelligent building design and management, leading to the growing importance of Building Digital Twins. This study focuses on functional, simulation-based digital twins that incorporate physics-based models for dynamic building performance simulation. The research investigates the interoperability of automatically generated Modelica models using TEASER (Tool for Energy Analysis and Simulation for Efficient Retrofit) and the AixLib library within the Modelon Impact simulation environment. TEASER was employed to generate reduced-order thermal building models from archetype data, which were then simulated using Modelon Impact, requiring a hybrid approach to address software compatibility issues. A comprehensive case study of a multi-level residential facility was conducted, evaluating thermal comfort and energy consumption at different levels of aggregation. The results demonstrate the feasibility of this integrated workflow for creating dynamic energy models, highlighting the trade-offs between modelling granularity and accuracy. The study contributes to more flexible and robust digital twin implementations for enhanced building performance analysis and operational decision-making.

Keywords: Building digital twin · Modelica · TEASER · AixLib · Simulation · Modelon

1 Introduction

The building sector significantly contributes to global energy consumption, accounting for 40% of European Union's energy use [1] and up to 35% of greenhouse gas emissions [2]. This underscores the need for sustainable, energy-efficient building practices and smarter design and management strategies [3]. In response, building energy performance simulation tools have advanced, becoming essential for predicting behavior, assessing risks, and complying with regulations. Among these tools, the Building Digital Twin (a digital replica of a building's physical and operational characteristics) has

© The Author(s) 2026
A. Jurelionis et al. (Eds.): BDTIC 2025, LNCE 775, pp. 13–25, 2026.
https://doi.org/10.1007/978-3-032-09040-9_2

gained increasing attention [4]. Depending on their purpose, building digital twins vary in form: *informational* twins (e.g., BIM systems or dashboards) focus on data visualization [5], while *functional* or simulation-based twins use physics-based models to simulate building performance dynamically, enabling predictive analysis and optimization [6]. This study focuses on the latter type, which supports operational decision-making and energy efficiency. Central to functional digital twins is physics-based simulation, especially through white-box modeling. Unlike black-box or grey-box models, white-box approaches rely on physical laws for accurate and interpretable behavior modeling. *Modelica*, an equation-based, object-oriented language, supports this method through modular, reusable components that allow detailed simulation of HVAC systems, control strategies, and building physics [7]. Compared to traditional simulation tools like EnergyPlus or ESP-r, Modelica offers more flexibility and transparency in capturing dynamic interactions [8]. Open-source libraries such as AixLib, IBPSA, and Buildings have accelerated Modelica's adoption in the building performance field. Tools like TEASER (Tool for Energy Analysis and Simulation for Efficient Retrofit) automate the creation of reduced-order thermal models compatible with Modelica libraries, using minimal input data [9]. This is particularly useful for urban energy modeling, where detailed building data is often unavailable. TEASER leverages typology databases and standards to streamline model generation, providing reasonable physical fidelity with limited control over fine details.

In comparison to other tools like IDEAS [10] and BuildSysPro [11], TEASER stands out for its open-source accessibility, integration with Modelica, and focus on scalable modeling for stock-level assessments. Once models are built, verification and scenario testing are necessary for validating performance. Platforms such as Modelon Impact, Dymola, and OpenModelica support these processes through the Functional Mockup Interface (FMI), which enables model exchange and co-simulation across different environments [12]. Modelon, beyond being a solver, functions as a validation environment, supporting control strategy integration, scenario testing, and real-time data coupling—critical for operational deployment of digital twins.

The TEASER + AixLib modeling workflow has been widely applied with Dymola. For instance, [9] used TEASER to generate reduced-order urban models simulated in Dymola. [13] showed scalable parameterization using archetype buildings, while [14] and [15] detailed AixLib's integration and structure for collaborative development in Dymola. Validation of high-order thermal models in AixLib has also been demonstrated in [16]. Despite these advancements, the interoperability of the TEASER + AixLib workflow within *Modelon Impact* has not been thoroughly explored. This study addresses that gap by evaluating the integration of TEASER-generated Modelica models with Modelon. A detailed case study is presented using real building data to create and validate a comprehensive digital twin. The model features multiple rooms, diverse occupancy profiles, and a detailed HVAC system, enabling realistic evaluation of thermal comfort and energy consumption. This work demonstrates the reproducibility and practical viability of the workflow in Modelon Impact and broadens the application of TEASER-based pipelines to new simulation contexts, contributing to more adaptable and robust digital twin solutions.

2   Method

The method implemented in this work utilizes TEASER and AixLib within the Modelica language, simulated through the Modelon Impact software environment.

2.1   Teaser

TEASER (Tool for Energy Analysis and Simulation for Efficient Retrofit) is a modeling tool designed to simplify energy analysis of buildings, with a focus on simulating thermal performance. It generates *reduced-order models* (ROMs) using RC (Resistance-Capacitance) networks, balancing reduced complexity with sufficient accuracy to allow faster energy simulations. Users provide key building data such as dimensions, orientation, and construction year, and define materials using TEASER's built-in or custom databases. The building is segmented into *thermal zones*, and HVAC systems, equipment parameters, occupancy schedules, and internal loads are specified. TEASER leverages integrated databases on materials, thermal properties, and climate to streamline this setup, automatically converting inputs into simplified thermal models through an RC-based approach. As shown in Fig. 1, the TEASER workflow begins with **data acquisition**, collecting essential building details like function, volume, and age.

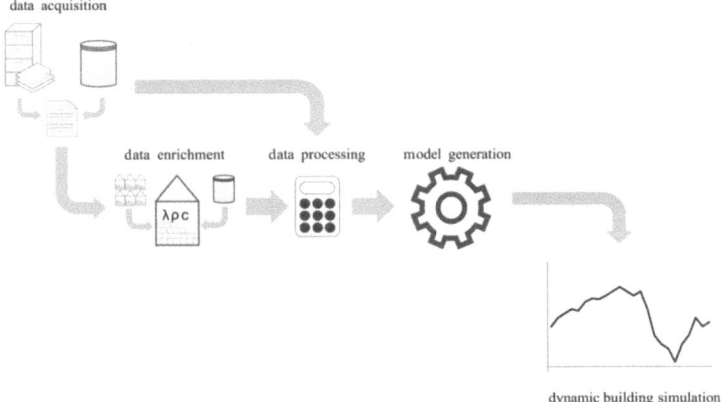

Fig. 1. TEASER workflow [9].

During **data enrichment**, this sparse input is completed using statistical norms and archetype libraries to create a detailed thermal and geometric profile. In **data processing**, the information is structured for simulation. Finally, **model generation** produces ROMs compatible with the Modelica-based *AixLib* library, supporting dynamic, scalable simulations. Thanks to its structured, automated approach, TEASER efficiently supports simulations ranging from individual buildings to large urban districts, requiring minimal manual input.

2.2 AixLib's Building Model

TEASER uses AixLib as the target library for generating simulation-ready ROMs in the Modelica language. It automates the creation of these models by mapping enriched building data (such as geometry, construction types, usage, and internal gains) onto predefined AixLib templates. TEASER calculates all necessary parameters and exports fully structured Modelica files that utilize AixLib components, enabling dynamic building performance simulations. The building model is generated using the *AixLib.ThermalZones.ReducedOrder.Multizone* framework, specifically targeting models like *MultizoneEquipped*, which support multiple thermal zones with internal gains and system interfaces. These models are based on standardized RC-network approaches, primarily following *VDI 6007–1* [17] and *DIN EN ISO 13790* [18], allowing for efficient yet physically meaningful dynamic simulations. TEASER automatically maps enriched building data onto these models and allows user-defined configurations, such as the number of zones and RC elements per wall, or whether to separate window resistances. Figure 2 shows the thermal zone module generated with TEASER using AixLib library.

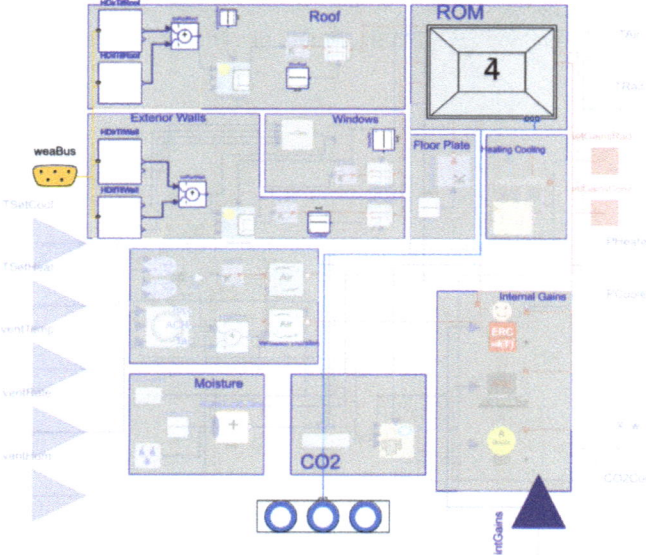

Fig. 2. Thermal zone module using TEASER and AixLib.

The ROMs are then imported into AixLib, where more detailed HVAC models, including the Generic AHU, are added. The Generic AHU model in AixLib includes detailed hydraulic systems for preheating, heating, and cooling, as well as a heat exchanger for heat recovery and a humidifier. The AHU is designed with supply and return air circuits, equipped with sensors and controlled by PID controllers to regulate airflow, temperature and humidity. Figure 3 represents the integration of the AHU model with the thermal zones model.

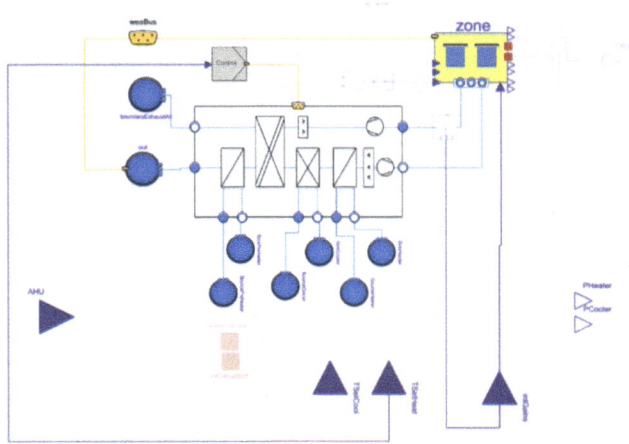

Fig. 3. Thermal zones with the AHU model in AixLib.

2.3 Modelon Impact as Simulation Platform

For this project, *Modelon Impact* was employed as the simulation environment [19]. Modelon Impact is a cloud-based modelling and simulation platform developed by Modelon, designed to facilitate efficient system-level simulation and optimization. It is built upon open Modelica language standards.

2.4 Compatibility Between TEASER, AixLib and Modelon

The Optimica Compiler Toolkit (OCT) serves as the computational engine behind Modelon Impact, incorporating both a compiler and a solver for executing Modelica-based simulations. OCT is compliant with the Modelica Language Specification (MLS) version 3.4 and supports the Modelica Standard Library (MSL) versions 3.2.3 and 4.0.0, with two key exceptions: Synchronous Language Elements (Chapter 16) and State Machines (Chapter 17) are not supported. As a result, any models that depend on features from these sections of the specification cannot be executed within the Modelon Impact environment, regardless of the selected MSL version.

The first officially released version of AixLib (v1.0.0) is compatible with Modelica 3.2.3; however, its implementation of the Air Handling Unit (AHU) model relies on constructs from both Synchronous Language Elements and State Machines. Consequently, this component cannot be simulated in Modelon Impact due to the aforementioned limitations of OCT. To resolve this, a later version of AixLib (v1.3.2, based on Modelica 4.0.0) was selected, as it includes a Generic AHU model that avoids the use of unsupported language features and is therefore compatible with Modelon Impact. However, integrating AixLib v1.3.2 introduced a new issue: the *ThermalZone* model in this version makes use of elements from the same unsupported chapters, thereby reintroducing compatibility problems during simulation. To overcome this, a hybrid approach was adopted: the Generic AHU model from AixLib v1.3.2 was manually integrated into the building model framework generated using AixLib v1.0.0, which retains a *Thermal-Zone* implementation free from unsupported features. The AHU was manually replaced and the necessary adjustments were made to ensure compatibility between the two versions. This solution enabled successful simulation of the building model (including the Generic AHU) within the Modelon Impact environment, while respecting the language constraints imposed by OCT. As a summary, Fig. 4 shows the compatibilities (linked through green arrows) and incompatibilities (with red arrows) between the different versions, and the final implementation.

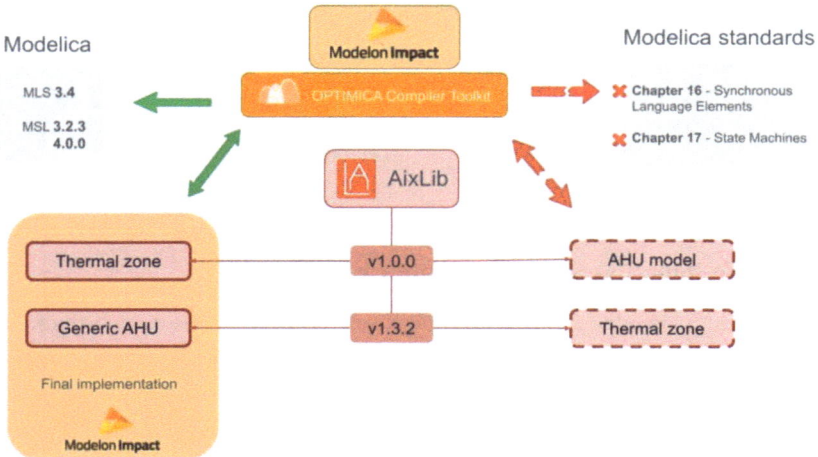

Fig. 4. Compatibility between Modelon Impact and AixLib.

2.5 Case Study

The case study building is a modular, multi-level residential facility located in Gothenburg, Sweden. It includes four floors and a rooftop with photovoltaic panels for on-site energy generation. With a total area of 1,518.3 m^2 and a volume of 5,040.2 m^3, the layout enables detailed spatial and thermal zoning. The first and third floors host experimental energy system modules, while the second and fourth floors contain residential units and shared spaces. The flexible internal structure supports various configurations for research or occupancy needs. Real-time sensors monitor indoor/outdoor temperature, humidity, CO_2 levels, energy consumption, and solar production. A centrally managed HVAC system includes a configurable Air Handling Unit (AHU) with heat recovery, preheating, cooling, humidification, and reheating. This modular and sensor-rich setup makes it ideal for validating dynamic energy models and control strategies. Building data, including geometry, materials, internal conditions, and systems, were introduced into the digital modeling workflow using TEASER.

Geometric Layout and Zoning. The four-story building was modeled based on architectural plans and divided into thermal zones according to use, control presence, and spatial function. Volumetric data, including floor area and ceiling height, was added per floor, maintaining the total volume of 5,040.2 m^3. Zones were connected to accurately reflect inter-zone heat transfer.

Construction Materials and Thermal Envelope. Material properties for walls, floors, ceilings, and windows were based on available documents or inferred using regional construction standards. Each envelope element was assigned thermal properties (U-values, conductivity, density, specific heat) using TEASER's material database.

Internal Heat Gains and Occupancy Profiles. Internal gains were set based on occupancy patterns and activity levels. Each zone had a detailed schedule aligned with its function—for example, bedrooms used a residential profile, while studios and shared spaces followed office-like usage per ASHRAE 90.1 [20]. A baseline metabolic heat gain of 70 W/person was adjusted for each space type. Lighting and equipment loads were defined using surface-specific power densities from ASHRAE standards. These loads were scheduled over 24-h cycles to simulate dynamic daily and weekly use, serving as time-dependent inputs for Modelica simulations in *Modelon Impact*.

Ventilation Flow Rate Calculation. Ventilation rates, unavailable from direct measurements, were estimated using DIN 1946 standards based on room function. Bedrooms included values for both sleeping areas and adjacent restrooms, while shared spaces received higher air change rates. These were applied in the AHU setup to reflect realistic airflow.

HVAC System Configuration. The central HVAC system was modeled in TEASER using a configurable AHU template. Components included heat recovery, preheater, cooler, humidifier, and reheater, plus independent fans. Temperature and humidity were controlled via setpoints with PID-regulated flow. The system supports both heating and cooling for ventilation and thermal comfort in each zone.

Simulation Setup and Zone Selection. Simulations aimed to assess thermal comfort and energy use across three aggregation levels: room, floor, and building. Three representative zones were selected: a west-facing private sleeping area (*Room*), a multipurpose meeting space (*Studio*), and a high-occupancy recreational area (*Multiroom*). These zones reflect diverse thermal behaviors. In addition to individual room simulations, zones were aggregated at floor and full-building levels, treating each as a single thermal zone for comparison.

The performance of the temperature control system is evaluated by measuring the thermal discomfort duration [tdd], defined as the percentage of the total simulation time in which the indoor temperature is outside the thermal comfort limits. It is computed according to Eq. (1).

$$tdd[\%] = \frac{\sum_{t=1}^{T} TimeT_z outside \left[T_z^{SP} - \Delta T, T_z^{SP} + \Delta T\right]}{T} \cdot 100 \qquad (1)$$

where T_z is the indoor zone temperature of the room, $T_z{}^{sp}$ is the temperature setpoint, ΔT is the deadband defined based on thermal comfort limits and T is the total simulation time. Two cases are evaluated: more strict thermal comfort limits ($\Delta T = 0.5°C$) and loosen thermal comfort bounds ($\Delta T = 1°C$). Additionally, the indoor temperature deviation with respect to the given setpoint is calculated through the normalized root-mean square deviation error NRMSD (Eq. (2)).

$$NRMSD(T_Z)[\%] = \frac{RMSD(T_z)}{T_z^{max} - T_z^{min}} = \frac{\sqrt{\frac{\sum_{t=1}^{T}\left(T_z^{sp} - T_z\right)^2}{T}}}{T_z^{max} - T_z^{min}} \cdot 100 \tag{2}$$

where RMSD is the root-mean square deviation error, T_Z^{max} and T_Z^{min} the maximum and minimum temperature in the interval respectively. The power consumption associated with each room or aggregated thermal zone is also estimated.

3 Results and Discussion

The building model generated at different aggregation levels is simulated for the period from May 18th to May 25th as it is a week with both heating and cooling demand for the AHU, and with a discretization of 1 h. A variable temperature setpoint profile is defined for the rooms in order to better evaluate the tracking capacity of the temperature control, ranging from 15 °C (for unoccupied times) to 21 °C (for occupied hours).

3.1 Room Level Simulations

The three selected spaces are simulated for the whole week, resulting in the temperature tracking performance shown in Table 1.

Table 1. Temperature tracking metrics at room simulation level.

Room name	$\Delta T = 0.5$°C		$\Delta T = 1$°C	
	tdd	NRMSD(Tz)	tdd	NRMSD(Tz)
Room	5.92%	3.31%	0.00%	0.00%
Studio	0.59%	2.89%	0.00%	0.00%
Multiroom	7.69%	3.50%	0.00%	0.00%

In stricter thermal comfort conditions, the thermal discomfort duration varies depending on the room. While thermal discomfort is more notable for *Room* and *Multiroom*, the *Studio* has a negligible discomfort duration. In all cases, the presented deviation in the temperature is low (few tenths of degrees). If more relaxed thermal comfort limits are considered, the temperature control can comply with those limits 100% of the time. Figure 5 shows in detail the temperature tracking for *Room*, where it can be observed that even if the temperature setpoint is highly variable during the week, the deviations in the indoor temperature are low.

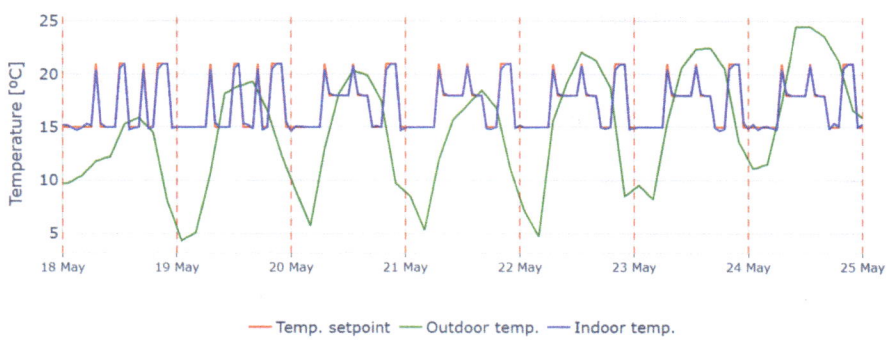

Fig. 5. Temperature tracking in Room: indoor, outdoor and setpoint.

Table 2 presents the consumption of each AHU component for each space. The results show a significant heating and cooling demand for the given week. Consumption is more noticeable in the *Multiroom* and *Studio* as they have higher surface, more occupants and with a use involving higher internal heat gains.

Table 2. Consumption of AHU at room level.

Room name	AHU component	Maximum power [W]	Total energy consumption [kWh]
Room	Preheater	65.27	3.85
	Heater	2355.58	187.92
	Cooler	2317.69	271.90
	Total	-	463.67
Studio	Preheater	59.64	3.84
	Heater	4721.22	522.14
	Cooler	4572.49	506.66
	Total	-	1032.64
Multiroom	Preheater	63.76	4.00
	Heater	5517.44	493.22
	Cooler	5766.89	677.38
	Total	-	1174.60

3.2 Floor and Building Level Simulations

Higher aggregation levels are simulated using the models generated by TEASER for the same boundary conditions. Table 3 presents the temperature tracking metrics for floor- and building-level simulations.

Table 3. Temperature tracking metrics at floor and building level.

Aggregation level	$\Delta T = 0.5\ °C$		$\Delta T = 1\ °C$	
	tdd	NRMSD (Tz)	tdd	NRMSD (Tz)
Floor	34.32%	18.77%	18.77%	9.37%
Building	60.95%	22.00%	36.69%	10.25%

In this case, the higher temperature tracking errors observed in floor-level and building-level simulations, compared to more accurate room-level simulations, are due to the necessity to generalize key variables across larger spatial scales. At the room level, detailed ventilation rates and occupancy patterns can be specifically defined, leading to minimal errors. However, when entire floors or the whole building are treated as single thermal zones, variables like ventilation flow and occupancy must be aggregated and are no longer tailored to individual spaces. This loss of specificity in representing factors that directly influence temperature results in a less faithful model of thermal dynamics and consequently, increased temperature tracking errors as the simulation encompasses larger, more generalized areas. Table 4 shows the corresponding consumptions for the whole floor and whole building.

Table 4. Consumption of AHU at floor and building level.

Room name	AHU component	Maximum power [W]	Total energy consumption [kWh]
Floor	Preheater	212.72	10.81
	Heater	27961.37	1553.46
	Cooler	16462.12	1197.31
	Total	-	2761.58
Building	Preheater	402.93	20.94
	Heater	42915.30	3843.58
	Cooler	23353.07	1593.43
	Total	-	163207.52

The simulations revealed that the level of model aggregation significantly impacts the accuracy of temperature tracking. Detailed modelling at the room level, which allows for the precise incorporation of ventilation rates and internal heat gain patterns, yields highly accurate results. However, as larger areas such as entire floors or the whole building are aggregated into single thermal zones, the necessity to generalize key variables leads to increased temperature tracking errors. Notably, the analysis also highlighted potential areas for improvement, such as optimizing insulation and HVAC systems to further reduce energy consumption. A key technical challenge encountered during the development process was ensuring compatibility between the different software versions of TEASER, AixLib, and Modelon Impact. The manual integration of the Generic AHU model from AixLib v1.3.2 with the building model framework generated using AixLib v1.0.0 was necessary to overcome limitations related to the Optimica Compiler Toolkit within Modelon Impact. This hybrid approach successfully enabled the simulation of the complete building model, including detailed HVAC components, within the chosen platform. The study demonstrates the practical feasibility and reproducibility of this workflow, extending the applicability of TEASER-based modelling pipelines into new simulation contexts like Modelon Impact.

4   Conclusions

This study successfully evaluated the interoperability of automatically generated Modelica models using the TEASER + AixLib workflow within the Modelon Impact simulation environment. The methodology implemented leveraged TEASER for the rapid generation of thermal building models based on archetypes and standardized input data, with these models constructed using the comprehensive AixLib Modelica library. The simulations were then executed using the robust Modelon Impact platform. The research demonstrated the reproducibility and practical feasibility of this integrated workflow, highlighting its potential for more flexible and robust digital twin implementations in building performance analysis.

The findings underscore the importance of modelling granularity, as room-level simulations exhibited superior accuracy in temperature tracking compared to aggregated floor and building-level models. Future research should focus on improving modelling tools and exploring advanced optimization strategies such as MPC. Additionally, the integration of real measurements to be used by the digital twin and advance its functionalities will be studied in future work.

Acknowledgements. This research was conducted in the framework of FEDECOM project, which has received funding from the European Union's Horizon Europe programme under Grant Agreement No. 101075660.

References

1. United Nations Environment Programme: 2022 Global Status Report for buildings and Construction: Towards a Zero-emission, Efficient and Resilient Buildings and Construction Sector. Nairobi (2022)
2. European Environment Agency: Greenhouse gas emissions from energy use in buildings in Europe 2022. https://www.eea.europa.eu/ims/greenhouse-gas-emissions-from-energy. Accessed 28 Feb 2023
3. Hafez, F.S., et al.: Energy efficiency in sustainable buildings: a systematic review. J. Build. Eng. **57**, 104945 (2022). https://doi.org/10.1016/j.esr.2022.101013
4. Yoon, S.: Building digital twinning: Data, information, and models. J. Build. Eng. **76**, 107021 (2023). https://doi.org/10.1016/j.jobe.2023.107021
5. Opoku, D.J., Perera, S., Osei-Kyei, R., Rashidi, M.: Digital twin application in the construction industry: a literature review. J. Build. Eng. **40**, 102726 (2021). https://doi.org/10.1016/j.jobe.2021.102726

6. Boje, C., Guerriero, A., Kubicki, S., Rezgui, Y.: Towards a semantic construction digital twin: directions for future research. Autom. Constr. **114**, 103179 (2020). https://doi.org/10.1016/j.autcon.2020.103179

7. Wetter, M., Zuo, W., Nouidui, T.S., Pang, X.: Modelica buildings library. J. Build. Perform. Simul. **4**(4), 1–16 (2011)

8. Coakley, D., Raftery, P., Keane, M.: A review of methods to match building energy simulation models to measured data. Renew. Sustain. Energy Rev. **37**, 123–141 (2014). https://doi.org/10.1016/j.rser.2014.05.007

9. Remmen, P., Lauster, M., Mans, M., Fuchs, M., Osterhage, T., Müller, D.: TEASER: an open tool for urban energy modelling of building stocks. J. Build. Perform. Simul. **11**(1), 84–98 (2017)

10. Baetens, R., De Coninck, R., Jorissen, F., Picard, D., Helsen, L., Saelens, D.: OPENIDEAS – An open framework for integrated district energy simulations. In: Proceedings of BS2015: 14th Conference of the International Building Performance Simulation Association, pp. 347–354. Hyderabad, India (2015)

11. Plessis, G., Kaemmerlen, A., Lindsay, A.: BuildSysPro: a Modelica library for modelling buildings and energy systems. In: Proceedings of the 10th International Modelica Conference, pp. 1161–1169. Lund, Sweden (2014). https://doi.org/10.3384/ecp140961161

12. Blochwitz, T., et al.: The Functional Mock-up Interface for Tool independent Exchange of Simulation Models. 8th International Modelica Conference, Dresden, Germany (2011)

13. Lauster, M., Mans, M., Remmen, P., Fuchs, M., Müller, D.: Scalable design-driven parameterization of reduced order models using archetype buildings with TEASER. In: BauSIM2016, pp. 535–542 (2016)

14. Maier, L., et al.: AixLib: an open-source Modelica library for compound building energy systems from component to district level with automated quality management. J. Build. Perform. Simul. **17**(2), 197–212 (2023)

15. Nytsch-Geusen, C., Rädler, J., Widl, E., Schweiger, G.: Structuring the building performance Modelica Library AixLib for open-source development. In: Proceedings of BS2015: 14th Conference of the International Building Performance Simulation Association (pp. [insert page numbers]). Hyderabad, India (2015)

16. Finkbeiner, K., et al.: Modeling a building energy system for development of energy management strategies in a shopping center. In: Proceedings of BS2017: 15th Conference of the International Building Performance Simulation Association, pp. 824–831. San Francisco, USA (2017)

17. VDI: VDI 6007-1: Calculation of transient thermal response of rooms and buildings – Modelling of rooms. Verein Deutscher Ingenieure, Düsseldorf (2015)

18. DIN: DIN EN ISO 13790:2008 – Energy performance of buildings: Calculation of energy use for space heating and cooling. Deutsches Institut für Normung, Berlin (2008)

19. Modelon Impact: https://modelon.com/modelon-impact/. Accessed 1 Apr 2025

20. ASHRAE, ANSI/ASHRAE Standard 90.1-2019: Energy Standard for Buildings Except Low-Rise Residential Buildings. American Society of Heating, Refrigerating and Air-Conditioning Engineers, Atlanta, GA (2019)

A Digital Twins Model Based on IFC Open BIM Models Managed on Web Platforms

Costantino Carlo Mastino[1](✉) ⓘ, Juozas Vaičiūnas[2] ⓘ, Raffaello Possidente[1] ⓘ,
Andrea Frattolillo[1] ⓘ, Mohsen Zavari[1] ⓘ, and Valerio Da Pos[3]

[1] D.I.C.A.AR.–University of Cagliari, Cagliari, Italy
mastino@unica.it
[2] Faculty of Civil Engineering and Architecture, Kaunas University of Technology, Kaunas,
Lithuania
[3] Cadline Software S.r.l, Padova, Italy

Abstract. The use of BIM has introduced a process of building digitalization in Europe and worldwide, including all the systems plant they contain. Italy and Lithuania, following European directives, have introduced national laws mandating the use of BIM for various projects. One of the challenges in using BIM platforms is ensuring data interoperability over the years. Currently, the only globally recognized model based on an open standard by ISO is the IFC data model. The management of buildings, with particular reference to energy and environmental aspects, is now one of the main objectives that all European Union states must pursue to ensure increasingly sustainable buildings. The use of digital models based on open BIM models could significantly contribute to the intelligent and environmentally sustainable management of buildings and systems. This work presents a case study of Digital Twins based on IFC open BIM models managed on a BIM web platform.

Keywords: Digital Twins · open BIM · IFC standard · IFC Web platforms · Energy managed

1 Introduction

The concept of DT was first proposed by Grieves (2005), who defined it as a virtual representation of a physical product with a bi-directional data flow. Later, Grieves (2014) expanded it into a system consisting of a physical product, a digital counterpart, and a two-way data link. Depending on the integration between physical and digital entities, Digital Model, Digital Shadow, and Digital Twin are distinguished [1]. The construction sector is witnessing a transformative shift towards digitization, driven by the need to enhance performance, ensure sustainability, and manage the inherent complexities of modern projects [2, 3].

By creating digital replicas of physical assets, construction projects can leverage Digital Twin models to monitor the condition of equipment and infrastructure, predict potential issues before they occur, and make data-driven decisions throughout the project

A. Jurelionis et al. (Eds.): BDTIC 2025, LNCE 775, pp. 26–39, 2026.
https://doi.org/10.1007/978-3-032-09040-9_3

lifecycle. This leads to enhanced collaboration, reduced errors, minimized downtime, and more precise project planning [4, 5].

Over the past decade, a notable evolution has been the transition from isolated digital tools to the more integrated framework of digital twins [6]. This framework brings together a myriad of technologies, including Building Information Modelling (BIM), Geographic Information System (GIS), Artificial Intelligence (AI), and real-time data sources like cameras, mobile devices, and sensors [7]. These technologies combine to facilitate the continuous exchange of information between the physical and digital realms. Driving and reflecting on this change, there is a body of research that underscores the potential of construction digital twins in improving decision-making processes, particularly during the production control phase of construction projects [6–13]. The DT concept comprises three elements: 1) a physical system, 2) a virtual system, and 3) a bidirectional data flow between the physical and virtual systems [8, 14]. Here, the virtual system should act as a digital replica or twin of the physical system, incorporating relevant data and simulation models. These models should be closely integrated, react to changes in the physical twin, and their granularity and accuracy should support the functional outputs or services delivered by the DT [15]. DT has the tendency to accelerate building energy efficiency by analyzing and monitoring a building's energy consumption. This data-driven approach aids in identifying inefficiencies and implementing strategies aimed at improving a facility's overall performance and lower energy consumption [16–18]. Table 1 examines the limitations and strengths in the application area as well as building energy efficiency. It also shows various digital technologies, such as DT, BIM, and Artificial intelligence (AI), that enhance building management and infrastructure, especially DT. The implementation of DT technologies in the construction sector has substantially increased efficiency and control. Smart sensors and building information management systems can monitor and control a construction site's operations [16, 18–21]. In construction works, DT uses sensors to continuously monitor a building's energy consumption, helping to reduce carbon emissions and improve the facility's quality [16, 18] Through the use of Heritage Building Information Modelling (HBIM) and BIM, DT can provide a comprehensive view of a building's energy consumption and perform accurate simulation and modelling, improving sustainability and lifecycle monitoring [17] Additionally, integrating IoT sensors into DT enables real-time energy monitoring, reducing carbon emissions and improving energy efficiency [22, 23]. According to Delval et al. (2024) [18], this technology can also be used to perform accurate monitoring without the need for regular inspections. For example, the Edge's Digital Twin integrates data from over 28,000 sensors monitoring temperature, humidity, occupancy, lighting, and more. This data feeds into a cloud-based platform, creating a digital replica of the building's systems, including HVAC, lighting, and energy consumption. The building achieved a 50% reduction in energy use compared to traditional offices, showcasing the benefits of real-time monitoring and predictive analytics [19]. Another, according to Yu et al. (2023) [24] to their discussion about how digital twin (DT) technology can significantly improve efficiency in port areas, mainly through the integration and optimization of renewable energy systems. It highlights that new energy power generation, such as renewable energy, is crucial for the low-carbon operation of ports. Also, Yu et al. (2023) [24] mentioned that the digital twin can provide more complete operation simulation capabilities, support interaction with various real factors, and gather operating experience to make performance evaluations more objective and accurate. This demonstrates the diverse applications of DT for enhancing energy efficiency in port areas. This category shows how Digital Twins enhances building energy efficiency by enabling real-time monitoring and predictive analytics, but high costs and integration challenges must be overcome for effective deployment.

Table 1. Strengths and limitations in the area of application (Building Energy Efficiency)

Categories	Strengths	limitations
Building energy efficiency	improved sustainability and management through renewable energy integration [25] Enhanced maintenance and comfort with DT and BIM [26] Real-time energy assessment and predictive accuracy [19] Collaboration support and early performance analysis with BIM [20]	Estimating carbon footprint and intervention cost [25] Challenges in integration into diverse practices [16] High costs and data quality.requirements [19] Technical constraints and setup costs [20]

The European Union promotes the adoption of Building Information Modeling (BIM) [27–32] to improve efficiency and transparency in the construction sector. Directive 2014/24/EU [33] on public procurement encourages member states to require the use of BIM in publicly funded projects. Many European countries have introduced the obligation of BIM for large public works. For example, the United Kingdom imposed Level 2 BIM from 2016 and has since made significant progress, while Germany, France, Italy, and Spain have adopted it gradually. In Italy, the BIM Decree (D.M. 560/2017) made BIM mandatory for contracts above certain economic thresholds from 2019, with progressive extension until 2025. In Lithuania, BIM adoption is also growing, with government support for the digitalization of construction. Since 2021, the use of BIM has been mandatory for certain public projects, aiming to improve efficiency, transparency, and sustainability in the construction sector. The goal of the European directive and various national implementations is to reduce waste, improve construction quality, and promote interoperability among industry professionals, accelerating the digital transition in the European construction sector. The evaluation of building performance is complex due to the engagements of multiple criteria during the design process. If the aim is to focus on demands like energy consumption, acoustic performance, thermal comfort, indoor air quality, and suchlike, evaluation should be correlated to the design process [34]. There are a variety of decision parameters when building's specifications are considered, including the envelope, heating, ventilating and air conditioning (HVAC) systems. The goal of the building performance can be set to reduce undesirable environmental influences while maximizing indoor air quality and energy efficiency [35]. In this regard, there is a huge need to obtain better building energy performance (BEP) without sacrifice of comfort, cost, aesthetics, or other performance considerations, and the application of different strategies and improvement of technologies for energy efficiency have been increasing dramatically [36]. The performance-based design process as explained by Kalay obtains qualitative solutions for specific unifications of forms and functions in particular conditions rather than process-based paradigms [37]. Moreover, these can only be detected with multi-criteria and multidisciplinary performance evaluations. For instance, the design process of existing net-zero energy buildings depends on performance-based decisions that contain "all aspects of passive building design, energy efficiency, daylight autonomy, comfort levels, renewable energy installations, [and] HVAC solutions, in addition to innovative solutions and technologies" [38]. To sum up, performance-based design evaluates a building's performance in respect of environmental considerations as well as design functions and aesthetics. This emphasizes the combination and extensive optimization of diverse measurable building performances [39]. Currently, building information modeling (BIM) provides a platform to incorporate various stages of the design process for the investigation of a building's performance. BIM is utilized not just as a model but also as a platform that includes all the characteristics of the building, and the disciplines and systems involved. It offers a suitable platform for co-working between multidisciplinary and interdisciplinary efforts during all processes of the project. Furthermore, because it conserves necessary information about energy performance analysis, when BIM is utilized for this, it can save considerable time and effort and reduce inconsistencies and mistakes [40]. Thus, this method becomes an encouraging way to obtain various design goals for architects and engineers [41]. An important consideration here is interoperability, expressed in terms of the ability of communication involving the exchange and usage of data among at least two software tools by a majority. There is no requirement for duplication of data with the help of interoperability when transferring data between software tools, while an ability to use multiple tools with the same sets of files for different aims is desired [42]. Interoperability enables data transition among applications and the collective contribution of multiple applications. Expressed, thus, as the capability of data exchange among applications helps to improve workflows and eliminates the need for the manual copying of data from previously created applications. Such copying limits the number and range of repetitions practically available to calculate best solutions for complicated subjects like energy design, and it also carries consistency issues [43]. Interoperability through open standards, particularly the IFC (Industry Foundation Classes) data schema [44, 45], has become pivotal in enabling high-level semantic information exchange across the AEC industry. As an open, standardized file format, IFC ensures seamless data sharing among stakeholders using diverse software applications, serving as a cornerstone of the BIM process by maintaining data integrity across platforms. This foundation supports Open BIM, a collaborative methodology that relies on interoperable formats like IFC to foster transparency and coordination among architects, engineers, contractors, and facility managers. Beyond design and construction, BIM's digital model evolves into a digital twin—a dynamic virtual replica enriched with real-time data to simulate and optimize the physical building's performance throughout its lifecycle. Complementing these advances, open scripting empowers algorithmic solutions, automating workflows and expanding customization possibilities across the AEC sector [46, 47]. Thanks to IoT sensors, BIM models, and advanced algorithms, digital twins allow monitoring, predictive analysis, and optimized maintenance of buildings and facilities. Integration through the use of Open BIM models and digital twins revolutionizes the construction sector, offering advantages that can be summarized in the following points:

- Interoperability between different software and platforms.
- Real-time monitoring of building performance.
- Predictive simulations to optimize consumption and maintenance. • Intelligent building management throughout its lifecycle.
- Use and integration of artificial intelligence in the management of building systems and facilities.

In the context of systems (HVAC, electrical, hydraulic), the digital twin allows the detection of anomalies, prevention of failures, and improvement of energy efficiency. Open BIM ensures that information is accessible to all involved professionals, reducing errors and operational costs. This synergy between open digital models and real-time data represents the future of smart building automation, improving sustainability from various perspectives (energy, environmental, economic, etc.) and the multi-objective comfort of built environments.

1.1 The HVAC Simulation Laboratory

In this work, the design of a prototype plant developed at the University of Cagliari in the Faculty of Engineering and Architecture is presented, which will allow the simulation of systems plant behavior in various internal climate configurations. Everything will be managed and monitored through a digital twin created using the systems OPEN BIM model with the use of a WEB OPEN BIM platform capable of operating on BIM models in open IFC format [48–50]. The prototype plant will be developed in the Laboratory of Technical Physics and Energy at the university and will have the capability to manage, always through the Digital Twin, on a BIM Web platform, as many as 14 different plant configurations. Figure 1 shows the different plant configurations planned for the prototype. As can be seen, various generation systems are provided that allow the simultaneous production of thermal and refrigeration energy. There are also thermal and photovoltaic solar panels that will enable a more accurate simulation of the plant size related to the climatic location and their producibility. The systems based on renewable energy production can be studied for different types of envelope. In fact, the laboratory will allow the simulation of different types of building components (walls, floors, roofs) through a dynamic simulation model, managed through the WEB platform, which will allow the simulation of multiple configurations to maximize energy efficiency and self-consumption of energy produced from renewable sources.

2   Components 'Open Bim for Hvac Test Facility Management'

A prototype has been developed consisting of innovative systems that combine air conditioning and electricity production, all managed remotely through advanced technology applied to the digital twin. This prototype uses sensors distributed throughout the system

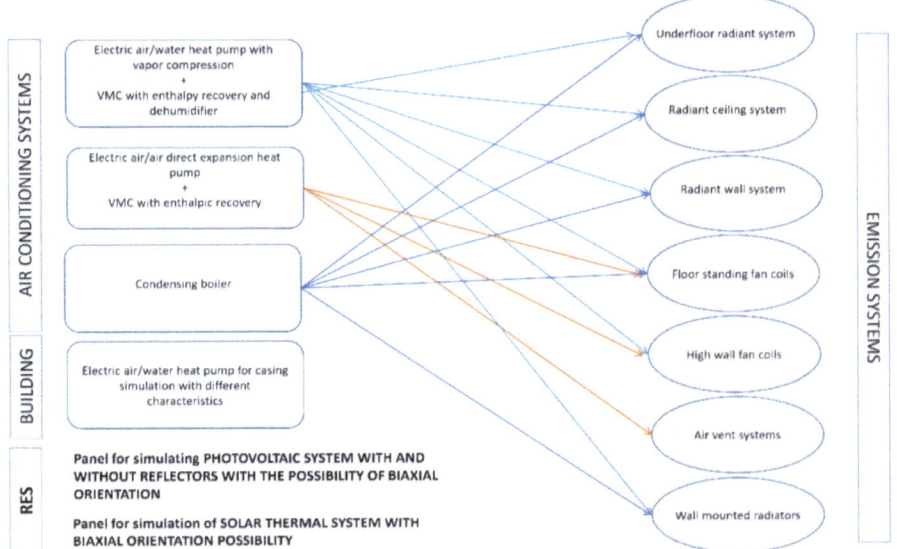

Fig. 1. A figure caption is always placed below the illustration. Short captions are centered, while long ones are justified. The macro button chooses the correct format automatically.

to collect real-time data on temperature, humidity, energy consumption, and equipment operating status. This data is sent to a digital model of the system, the digital twin based on OPEN BIM, which simulates the behavior of the physical system. The digital twin enables monitoring and optimization of system performance and allows for automatic adjustment of operating parameters to maximize energy efficiency and environmental comfort. Additionally, the system can predict faults and/or malfunctions and suggest preventive maintenance interventions, with the capability to integrate artificial intelligence. The integration of electricity production, for example through photovoltaic solar panels, reduces dependence on the electrical grid and lowers energy costs. Remote management offers the ability to control the system from anywhere using connected devices such as smartphones or tablets. This advanced system not only improves operational efficiency but also contributes to environmental sustainability by reducing CO_2 emissions and promoting the use of renewable energy while maximizing self-consumption. In summary, this prototype represents a step forward toward smarter, more sustainable, and autonomous buildings.

2.1   Air Conditioning System

The air conditioning systems designed for the prototype are based on various types of generators, including heat pump generators with direct expansion, air-to-water systems, absorption units, and, finally, a small combustion generator. For example, Mitsubishi mono split air conditioners (with direct expansion and air-cooled condensers) are known for their efficiency, reliability, and advanced technology. These air conditioning systems are designed to provide optimal comfort in residential and commercial environments, ensuring precise temperature control and quiet operation.

2.2 CMV Air Exchange System

Controlled Mechanical Ventilation (CMV) systems with a variable air flow rate of up to 500 m^3/h and a thermodynamic heat recovery unit represent an advanced solution for ensuring high-quality indoor air and optimal energy efficiency. These systems are ideal for medium-sized residential and commercial buildings, where maintaining constant and controlled air exchange is essential. The thermodynamic heat recovery unit is one of the distinctive features of these systems. This device not only recovers heat from the extracted air but also employs a thermodynamic cycle to further enhance the efficiency of heat transfer. This process allows the incoming air to be heated or cooled with minimal energy consumption, significantly reducing operating costs. Air quality is ensured by advanced filters that remove dust, pollen, and other pollutants, thus improving the healthiness of indoor environments.

2.3 Management System Based on Digital IFC Open BIM WEB Models

The Management System based on IFC Open BIM WEB BIM Leader digital models represents a revolution in the field of construction design and management. This system leverages Open BIM (Building Information Modeling) technology to create detailed and interoperable digital models using the IFC format. The adoption of IFC Open BIM enables efficient collaboration among all professionals involved in the project, such as architects, engineers, and builders, facilitating the exchange of information and reducing communication errors. Thanks to this interoperability, the digital models can be utilized throughout various phases of the building lifecycle, from design and construction to management and maintenance. The WEB BIM Leader system integrates these technologies into a web platform accessible from any internet-connected device. This allows for centralized and real-time project management, enhancing transparency and operational efficiency. Users can view, edit, and share digital models, monitor project progress, and coordinate activities more effectively. Another significant advantage of this system is the capability to perform advanced analyses and simulations, such as evaluating energy performance and cost management. This helps make informed decisions and optimize resources, contributing to more sustainable and profitable projects. In summary, the Management System based on IFC Open BIM WEB BIM Leader digital models is an innovative and comprehensive solution for managing construction projects, improving collaboration, efficiency, and sustainability. Figure 2 shows the main open formats for files managed [44, 45, 51, 52] through the BIMLeader platform created by BuildingS-MART International [45]. These formats allow the management of almost all necessary information through their open format, which enables information interoperability.

Figure 3 shows the developed prototype system consisting of the chamber equipped with various thermal emission systems (R1), a cold chamber capable of reaching very low temperatures (-26 °C), and all the systems for generating thermal and refrigeration energy. The 3D MEP model represented in Fig. 3 is a phase of project study and is currently under further study to be improved before its construction.

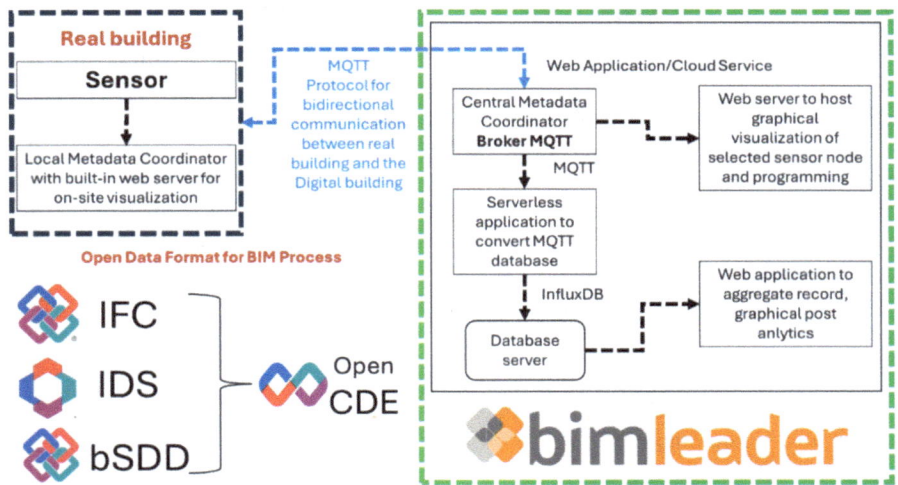

Fig. 2. Flow diagram of the Management System based on IFC Open BIM WEB digital models.

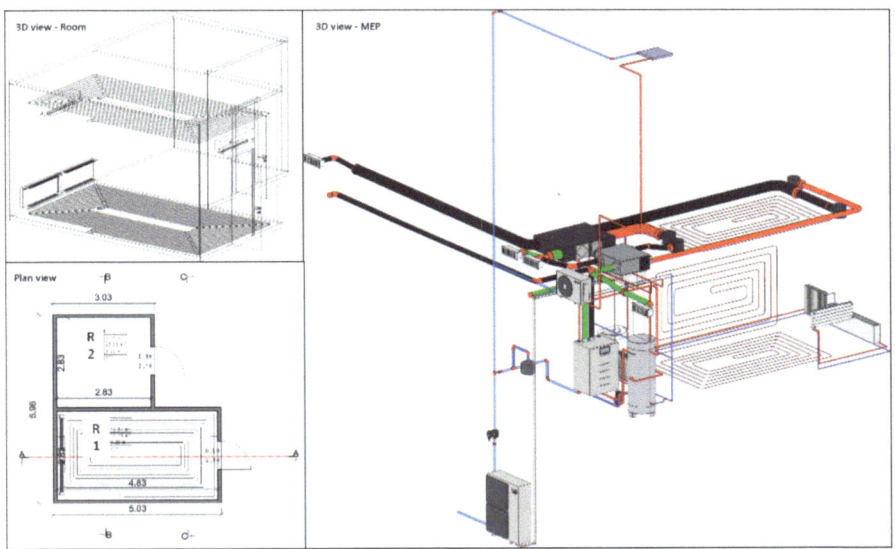

Fig. 3. Prototype system in the design phases: chamber and plant systems for the chamber.

3 Digital Model and MQTT Protocol

MQTT stands for Message Queuing Telemetry Transport and is a standard messaging protocol according to ISO/IEC 20922:2016. It is designed for communications requiring low energy consumption or when bandwidth is limited.

MQTT operates on a "publisher/subscriber" paradigm: every client can act as both a publisher and a subscriber, and it receives information only from the topics it has chosen to subscribe to. The MQTT broker (server) manages communication between connected devices. A subscriber client that wishes to receive information subscribes to the designated topic. When a publisher client has an update on one of its topics, it publishes it to the broker, which then sends the update to all subscriber clients registered to that topic. Efficient communication via MQTT offers numerous advantages both in terms of data handling and energy consumption, thereby benefiting the environment. This enables IIoT devices without continuous power supply to better utilize their batteries, optimize energy consumption, and extend their performance lifespan. Additionally, MQTT ensures reliable message delivery, which is particularly important in the context of industrial automation and process digitalization. Data acquisition and flow communication are essential aspects for maximizing benefits. In light of this, we can state that MQTT stands out for:

- Lightweight and reliable performance
- Scalability of communication
- Data security

3.1 Integration of KNX and MQTT

It is possible to use both protocols simultaneously, and in many cases, this combination can offer significant advantages. Indeed, various product series allow the integration of a KNX/MQTT network. They enable the publication and retrieval of data to/from an MQTT server and a KNX network. These gateways provide quick and easy access to the IoT world and are compatible with IoT servers supporting the MQTT protocol. MQTT communication can be encrypted using TLS/SSL protocols, ensuring a secure and protected connection. These converters are very easy to configure and enable the IoT system to interface with the KNX world. On the KNX side, it is possible to connect all common KNX devices such as temperature sensors, shutters, light switches, actuators, alarms, and more.

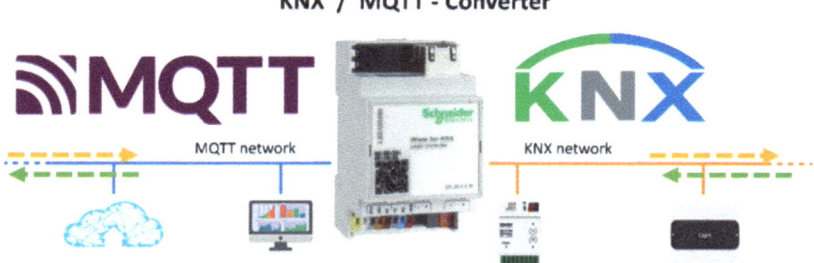

Fig. 4. Flow diagram of data exchange between prototype systems and the internet network

Figure 4 shows the interface between the KNX protocol, used for the management and control of all plant devices, and the MQTT protocol used for internet communication.

MQTT is a versatile and powerful protocol for communication between IoT devices. Its lightweight architecture and quality of service mechanisms make it ideal for applications where bandwidth and reliability are crucial. Therefore, its use is well-suited for the prototype under investigation in this research. In conclusion, the integration of the MQTT protocol into an open BIM web management platform for a prototype of air conditioning and electricity production systems represents a significant innovation. This combination enables efficient and secure communication between the various components of the system and the BIM model, facilitating real-time management and monitoring. The open BIM platform allows for greater interoperability and collaboration among different stakeholders, improving the design and maintenance of the system. MQTT ensures fast and reliable data transmission, optimizing performance and reducing response times. In summary, this integrated solution makes the system smarter, more efficient, and more sustainable, flexibly addressing energy and environmental needs.

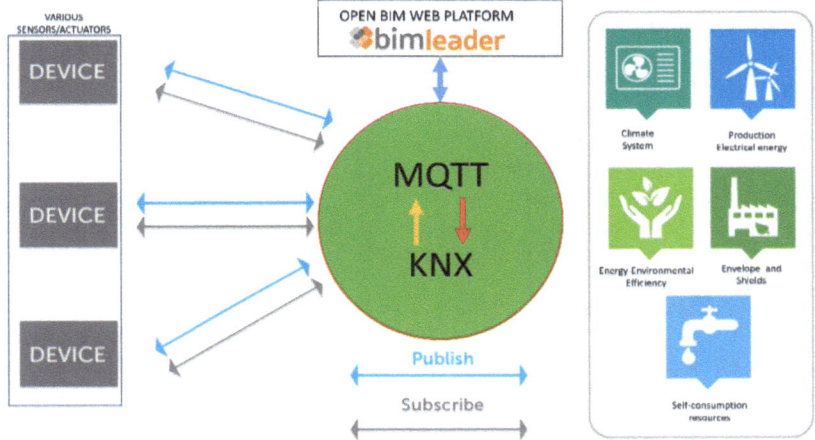

Fig. 5. Flow Diagram: MQTT–KNX and BIM Leader Web Platform

Figure 5 shows the flow diagram used to manage the different components of the prototype system through the MQTT and KNX protocols. These protocols will form the basis of bidirectional communication between the prototype system and its digital model.

4 Conclusion

In conclusion, a prototype has been studied to simulate and manage the building-plant system to be developed at the Engineering and Architecture campus of the University of Cagliari. The prototypal system will be created in the Technical Physics Laboratory and will have the ability to manage, always through the Digital Twin on a BIM Web platform, as many as 14 different plant configurations, significantly expanding the results obtainable in the field compared to the single specific configuration already adopted.

The goal is to create and develop a simulation, prediction, and management system for building-plant systems based on Open BIM technology managed on web platforms. This system will allow the implementation of management logic aimed at optimizing energy efficiency and maximizing self-consumption of energy produced by renewable energy sources, considering the various configurations of both the plant systems and the different types of building envelopes. The management system is a fundamental aspect of all plant systems and is designed based on a digital model replicated on a web platform that processes interoperable data formatted according to the Open BIM-IFC model. The use of open models based on BIM to realize the digital model for real-time management of the building-plant system represents a significant technological advancement. The entire initiative is aimed at maximizing energy and environmental efficiency, as well as self-consumption of energy produced by renewable energy sources (RES). This is achievable because the model can monitor and implement logic that considers not only the plant system but also the entire building-plant system within a specific environmental context in which it is actually inserted. The implementation of the prototype thus becomes a main and innovative aspect of this research.

References

1. Kritzinger, W., Karner, M., Traar, G., Henjes, J., Sihn, W.: Digital twin in manufacturing: a categorical literature review and classification. IFAC-Pap. **51**, 1016–1022 (2018). https://doi.org/10.1016/j.ifacol.2018.08.474
2. Murtagh, N., Scott, L., Fan, J.: Sustainable and resilient construction: current status and future challenges. J. Clean. Prod. **268**, 122264 (2020). https://doi.org/10.1016/j.jclepro.2020.122264
3. The next Normal in Construction (2020)
4. Moshood, T.D., Rotimi, J.Ob., Shahzad, W., Bamgbade, J.A.: Infrastructure digital twin technology: a new paradigm for future construction industry. Technol. Soc. **77**, 102519 (2024). https://doi.org/10.1016/j.techsoc.2024.102519
5. Magomadov, V.S.: The digital twin technology and its role in manufacturing. IOP Conf. Ser. Mater. Sci. Eng. **862**, 032080 (2020). https://doi.org/10.1088/1757-899X/862/3/032080
6. AlBalkhy, W., Karmaoui, D., Ducoulombier, L., Lafhaj, Z., Linner, T.: Digital twins in the built environment: definition, applications, and challenges. Autom. Constr. **162**, 105368 (2024). https://doi.org/10.1016/j.autcon.2024.105368
7. Hwang, B.-G., Ngo, J., Her, P.W.Y.: Integrated digital delivery: implementation status and project performance in the singapore construction industry. J. Clean. Prod. **262**, 121396 (2020). https://doi.org/10.1016/j.jclepro.2020.121396
8. Boje, C., Guerriero, A., Kubicki, S., Rezgui, Y.: Towards a semantic construction digital twin: directions for future research. Autom. Constr. **114**, 103179 (2020). https://doi.org/10.1016/j.autcon.2020.103179
9. Sacks, R., Girolami, M., Brilakis, I.: Building information modelling, artificial intelligence and construction tech. Dev. Built Environ. **4**, 100011 (2020). https://doi.org/10.1016/j.dibe.2020.100011
10. Tuhaise, V.V., Tah, J.H.M., Abanda, F.H.: Technologies for digital twin applications in construction. Autom. Constr. **152**, 104931 (2023). https://doi.org/10.1016/j.autcon.2023.104931
11. Saini, G.S., Fallah, A., Ashok, P., Van Oort, E.: Digital twins for real-time scenario analysis during well construction operations. Energies **15**, 6584 (2022). https://doi.org/10.3390/en15186584

12. Su, S., Zhong, R.Y., Jiang, Y., Song, J., Fu, Y., Cao, H.: Digital twin and its potential applications in construction industry: state-of-art review and a conceptual framework. Adv. Eng. Inform. **57**, 102030 (2023). https://doi.org/10.1016/j.aei.2023.102030

13. Long, W., Bao, Z., Chen, K., Thomas Ng, S., Yahaya Wuni, I.: Developing an integrative framework for digital twin applications in the building construction industry: a systematic literature review. Adv. Eng. Inform. **59**, 102346 (2024). https://doi.org/10.1016/j.aei.2023.102346

14. Grieves, M.: Digital twin: manufacturing excellence through virtual factory replication (2015)

15. Boyes, H., Watson, T.: Digital twins: an analysis framework and open issues. Comput. Ind. **143**, 103763 (2022). https://doi.org/10.1016/j.compind.2022.103763

16. Daniotti, B., et al.: The development of a BIM-Based interoperable toolkit for efficient renovation in buildings: from BIM to digital twin. Buildings **12**, 231 (2022). https://doi.org/10.3390/buildings12020231

17. Massafra, A., Predari, G., Gulli, R.: Towards digital twin driven cultural heritage management: a hbim-based workflow for energy improvement of modern buildings. Int. Arch. Photogramm. Remote Sens. Spat. Inf. Sci. XLVI-5/W1-2022, 149–157 (2022). https://doi.org/10.5194/isprs-archives-XLVI-5-W1-2022-149-2022

18. Delval, T., Rezoug, M., Tual, M., Fathy, Y., Mege, R.: Towards a digital twin system design based on a user-centered approach to improve quality control on construction sites. In Proceedings of the Advances in Information Technology in Civil and Building Engineering. Skatulla, S., Beushausen, H. (eds.) Springer International Publishing: Cham, pp. 579–596 (2024)

19. Tahmasebinia, F., Lin, L., Wu, S., Kang, Y., Sepasgozar, S.: Exploring the benefits and limitations of digital twin technology in building energy. Appl. Sci. **13**, 8814 (2023). https://doi.org/10.3390/app13158814

20. Vite, C., Horvath, A.-S., Neff, G., Møller, N.L.H.: Bringing human-centredness to technologies for buildings: an agenda for linking new types of data to the challenge of sustainability. In: Proceedings of the CHItaly 2021: 14th Biannual Conference of the Italian SIGCHI Chapter, ACM, Bolzano Italy, pp. 1–8 (2021)

21. Spudys, P., Afxentiou, N., Georgali, P.-Z., Klumbyte, E., Jurelionis, A., Fokaides, P.: Classifying the operational energy performance of buildings with the use of digital twins. Energy Build. **290**, 113106 (2023). https://doi.org/10.1016/j.enbuild.2023.113106

22. Ahmad, T., Zhang, D.: Using the internet of things in smart energy systems and networks. Sustain. Cities Soc. **68**, 102783 (2021). https://doi.org/10.1016/j.scs.2021.102783

23. Ferdaus, M.M., Dam, T., Anavatti, S., Das, S.: Digital technologies for a net-zero energy future: a comprehensive review. Renew. Sustain. Energy Rev. **202**, 114681 (2024). https://doi.org/10.1016/j.rser.2024.114681

24. Yu, P., Zhaoyu, W., Yifen, G., Nengling, T., Jun, W.: Application prospect and key technologies of digital twin technology in the integrated port energy system. Front. Energy Res. **10**, 1044978 (2023). https://doi.org/10.3389/fenrg.2022.1044978

25. Agostinelli, S., Cumo, F., Nezhad, M.M., Orsini, G., Piras, G.: Renewable energy system controlled by open-source tools and digital twin model: zero energy port area in Italy. Energies **2022**, 15 (1817). https://doi.org/10.3390/en15051817

26. Hosamo, H.H., Nielsen, H.K., Kraniotis, D., Svennevig, P.R., Svidt, K.: Improving building occupant comfort through a digital twin approach: a bayesian network model and predictive maintenance method. Energy Build. **288**, 112992 (2023). https://doi.org/10.1016/j.enbuild.2023.112992

27. ISO ISO 19650-1: Organization and digitization of information about buildings and civil engineering works, including building information modelling (BIM)—information management using building information modellingPart 1: concepts and principles (2018)

28. ISO 19650-2: Organization and digitization of information about buildings and civil engineering works, including building information modelling (BIM)—information management using building information modellingPart 2: delivery phase of the assets (2018)
29. ISO 19650–3: Organization and digitization of information about buildings and civil engineering works, including building information modelling (bim) — information management using building information modellingPart 3: operational phase of the assets (2020)
30. ISO 19650-5: Organization and digitization of information about buildings and civil engineering works, including building information modelling (BIM)—information management using building information modellingPart 5: security-minded approach to information management (2020)
31. ISO 19650–4: Organization and digitization of information about buildings and civil engineering works, including building information modelling (BIM)—information management using building information modellingPart 4: information exchange (2022)
32. ISO 19650-6: Organization and digitization of information about buildings and civil engineering works, including building information modelling (BIM)—information management using building information modellingPart 6: health and safety information (2025)
33. Directive 2014/24/EU of the European Parliament and of the Council of 26 February 2014 on Public Procurement and Repealing Directive 2004/18/EC Text with EEA Relevance (2014)
34. Hopfe, C.J., Augenbroe, G.L.M., Hensen, J.L.M.: Multi-criteria decision making under uncertainty in building performance assessment. Build. Environ. **69**, 81–90 (2013). https://doi.org/10.1016/j.buildenv.2013.07.019
35. Osmo Palonen, M., Hamdy, M., Hasan, A.: Mobo a new software for multi-objective building performance optimization (2013)
36. Petersen, S., Svendsen, S.: Method and simulation program informed decisions in the early stages of building design. Energy Build. **42**, 1113–1119 (2010). https://doi.org/10.1016/j.enbuild.2010.02.002
37. Kalay, Y.E.: Performance-based design. Autom. Constr. **8**, 395–409 (1999). https://doi.org/10.1016/S0926-5805(98)00086-7
38. Attia, S., Gratia, E., De Herde, A., Hensen, J.L.M.: Simulation-based decision support tool for early stages of zero-energy building design. Energy Build. **49**, 2–15 (2012). https://doi.org/10.1016/j.enbuild.2012.01.028
39. Shi, X., Yang, W.: Performance-driven architectural design and optimization technique from a perspective of architects. Autom. Constr. **32**, 125–135 (2013). https://doi.org/10.1016/j.autcon.2013.01.015
40. Rahmani Asl, M., Zarrinmehr, S., Yan, W.: Towards BIM-based parametric building energy performance optimization. Cambridge (Ontario), Canada, pp. 101–108 (2013)
41. Nguyen, A.-T., Reiter, S., Rigo, P.: A review on simulation-based optimization methods applied to building performance analysis. Appl. Energy **113**, 1043–1058 (2014). https://doi.org/10.1016/j.apenergy.2013.08.061
42. Sanhudo, L., Ramos, N.M.M., Poças Martins, J., Almeida, R.M.S.F., Barreira, E., Simões, M.L., Cardoso, V.: Building information modeling for energy retrofitting–a review. Renew. Sustain. Energy Rev. **89**, 249–260 (2018) https://doi.org/10.1016/j.rser.2018.03.064
43. Eastman, C.M.: BIM handbook: a guide to building information modeling for owners, managers, designers, engineers and contractors. John Wiley & Sons (2011). ISBN 0-470-54137-7
44. ISO 16739-1:2018 Industry Foundation Classes (IFC) for Data Sharing in the Construction and Facility Management Industries -- Part 1: Data Schema (2018)
45. buildingSMART Specification. http://Www.Buildingsmart-Tech.Org/Specifications
46. Flores, D.A.N., Guimarães, D.F.G.: Programa de pós-graduação em construção civil

47. Lilis, G.N., Wang, M., Katsigarakis, K., Mavrokapnidis, D., Korolija, I., Dimitrios, R.: BIM-based semantic enrichment and knowledge graph generation via geometric relation checking. Autom. Constr. **173**, 106081 (2025). https://doi.org/10.1016/j.autcon.2025.106081

48. Mastino, C.C., Baccoli, R., Frattolillo, A., Marini, M., Bella, A.D.: The building information model and the IFC standard: analysis of the characteristics necessary for the acoustic and energy simulation of buildings. In: Proceedings of the 3rd IBPSA-Italy conference Bozen-Bolzano; bu,press - Bozen-Bolzano University Press Free University of Bozen-Bolzano: Bozen-Bolzano, pp. 479–486 (2017)

49. Mastino, C.C., Baccoli, R., Frattolillo, A., Marini, M., Bella, A.D.: The building information model and the IFC standard: analysis of the characteristics necessary for the acoustic and energy simulation of buildings

50. Marini, M., Mastino, C.C., Baccoli, R., Frattolillo, A.: BIM and plant systems: a specific assessment. Energy Procedia **148**, 623–630 (2018). https://doi.org/10.1016/j.egypro.2018.08.150

51. BuildingSMART International Information Delivery Specification IDS (2023). https://Technical.Buildingsmart.Org/Projects/Information-Delivery-Specification-Ids/

52. BuildingSMART International buildingSMART Data Dictionary (bSDD). https://Www.Buildingsmart.Org/Users/Services/Buildingsmart-Data-Dictionary/

Overview of the Use of AI in Buildings Sustainability Assessment

Turkay Ersener[1] and Paris A. Fokaides[1,2(✉)] (ID)

[1] School of Engineering, Frederick University, Limassol, Cyprus
paris.fokaides@ktu.lt
[2] Faculty of Civil Engineering and Architecture, Kaunas University of Technology, Kaunas, Lithuania

Abstract. This paper presents an overview of how Artificial Intelligence (AI) supports the sustainability assessment of buildings, structured around the main phases of the building lifecycle. The review focuses on four core stages: design and planning, operation and monitoring, assessment and optimization, and compliance and certification support. Within this structure, the study highlights the growing role of AI in enhancing decision-making, improving building performance, and supporting sustainable outcomes across each phase. AI tools are categorized into four main groups: general-purpose platforms, data analytics environments, building-specific tools, and specialized applications. These are then mapped to key application domains such as prediction, simulation, decision support, and system optimization. The study emphasizes the connection between AI functionalities and specific needs in building sustainability, offering a structured approach for understanding the current landscape of tools and methods. By combining a lifecycle-based perspective with a classification of AI technologies and use cases, the paper aims to support researchers and practitioners in navigating the evolving intersection of AI and sustainable built environments. The findings serve as a foundation for further research and tool development, fostering more effective integration of AI in future sustainability assessments.

Keywords: Artificial Intelligence · building sustainability · building lifecycle · smart buildings · energy efficiency · decision support

1 Introduction

The building sector plays a central role in the global sustainability agenda, accounting for a significant share of energy consumption, carbon emissions, and material use. In response to increasing environmental and regulatory pressures, there is growing interest in tools and methodologies that can enhance the design, operation, and assessment of sustainable buildings. Among these, Artificial Intelligence (AI) is emerging as a powerful enabler, offering new possibilities for data-driven insights, predictive modeling, and performance optimization.

Recent advances in AI, including machine learning, pattern recognition, and data analytics, have opened up opportunities to address complex sustainability challenges in

© The Author(s) 2026
A. Jurelionis et al. (Eds.): BDTIC 2025, LNCE 775, pp. 40–49, 2026.
https://doi.org/10.1007/978-3-032-09040-9_4

the built environment. These range from optimizing energy performance and monitoring occupant behavior to supporting certification processes and informing policy compliance. However, while many studies explore AI from a technical standpoint, there remains a need for a structured understanding of how AI tools align with specific phases in the building lifecycle.

This study aims to bridge that gap by providing an overview of AI applications in building sustainability assessment, organized around four key lifecycle phases: design and planning, operation and monitoring, assessment and optimization, and compliance and certification support. Through this lens, the paper identifies tool categories, application domains, and emerging trends relevant to both researchers and practitioners.

2 Methodology

This study follows a systematic literature review approach, structured according to the PRISMA (Preferred Reporting Items for Systematic Reviews and Meta-Analyses) framework. The objective was to identify and analyze recent research exploring the application of Artificial Intelligence in the sustainability assessment of buildings. The literature search was conducted using the Scopus database, covering the period from January 2022 to March 2025. The search string included the keywords "Artificial Intelligence" and "Building Sustainability," which retrieved a total of 507 documents.

Table 1. Summary of PRISMA Selection Process.

Parameter	Value
Timeframe	January 2022 – March 2025
Database	Scopus
Search Keywords	"Artificial Intelligence" AND "Building Sustainability"
Initial Records Retrieved	507
Final Studies Analyzed	25
Review Framework	PRISMA
Analysis Tool	Biblioshiny (Bibliometrix R package)

A screening process was applied to identify studies specifically addressing the intersection of AI and building sustainability. After reviewing titles, abstracts, and keywords, a targeted subset of 25 studies was selected for detailed analysis. These studies were chosen based on their direct relevance to the topic, their methodological contribution, and their alignment with one or more phases of the building lifecycle. The bibliometric analysis of the selected literature was performed using Biblioshiny, the web-based interface of the Bibliometrix R package. Biblioshiny enabled the exploration of publication trends, identification of the most relevant sources, analysis of citation patterns across countries, and mapping of keyword co-occurrences and international collaborations. The results are presented through a series of visual outputs, including source impact graphs, word clouds, co-occurrence networks, and collaboration maps, offering a comprehensive picture of the research landscape. The key parameters of the review process and bibliometric analysis are summarized in Table 1.

3   Results and Discussion

3.1   Bibliometric Analysis of Literature Resources

Figure 1 presents a word cloud showing the most frequent terms in studies on AI and building sustainability. The word cloud chart visualizes the most frequently occurring terms in the selected literature on Artificial Intelligence and building sustainability. Dominant among them are "energy efficiency" and "artificial intelligence," reflecting the central focus of the field on using AI techniques to optimize energy performance in buildings. The prominence of related terms such as "energy utilization," "energy management," and "optimization" underscores the practical application of AI tools in improving operational performance, reducing consumption, and advancing sustainable development goals. The strong presence of "machine learning," "deep learning," and "decision making" indicates a methodological trend in the analyzed studies, suggesting that data-driven, predictive approaches are commonly employed to inform design and operational strategies. Additionally, terms like "forecasting," "neural networks," and "intelligent buildings" further point to the integration of AI into real-time systems and smart environments. Notably, keywords such as "sustainable development," "climate change," and "Internet of Things" illustrate the broader context in which this research is situated, linking building sustainability with global environmental and technological challenges. This chart thus provides a snapshot of the conceptual landscape and thematic emphasis within the field, highlighting the convergence of AI, energy-focused objectives, and the digital transformation of the built environment.

Fig. 1. Word cloud showing the most frequent terms in studies on AI and building sustainability (2022–2025).

The co-occurrence network chart (Fig. 2) maps the thematic relationships between keywords in the selected literature on Artificial Intelligence and building sustainability. Two distinct clusters are visible: the red cluster, which centers around "energy efficiency" and "artificial intelligence," and the blue cluster, which gravitates around terms related to smart energy infrastructure and digital technologies, such as "smart grid," "deep learning," and "internet of things." The prominence of "energy efficiency" and "artificial intelligence" as central nodes reflects their foundational role in this research domain. Closely linked keywords in the red cluster—such as "optimization," "decision making," "machine learning," and "sustainable development"—highlight the breadth of AI's applications in improving building performance, informing policy, and supporting climate-related goals. In contrast, the blue cluster reflects a more infrastructure- and systems-oriented focus, with strong links among terms like "energy management systems," "smart power grids," and "intelligent buildings." The connections between the two clusters suggest an interdisciplinary convergence, where AI-driven optimization and control strategies are increasingly integrated with smart energy systems and urban infrastructure. This network provides insight into the evolving structure of research topics, revealing how sustainability goals in buildings are being addressed through a combination of intelligent technologies and system-wide innovations.

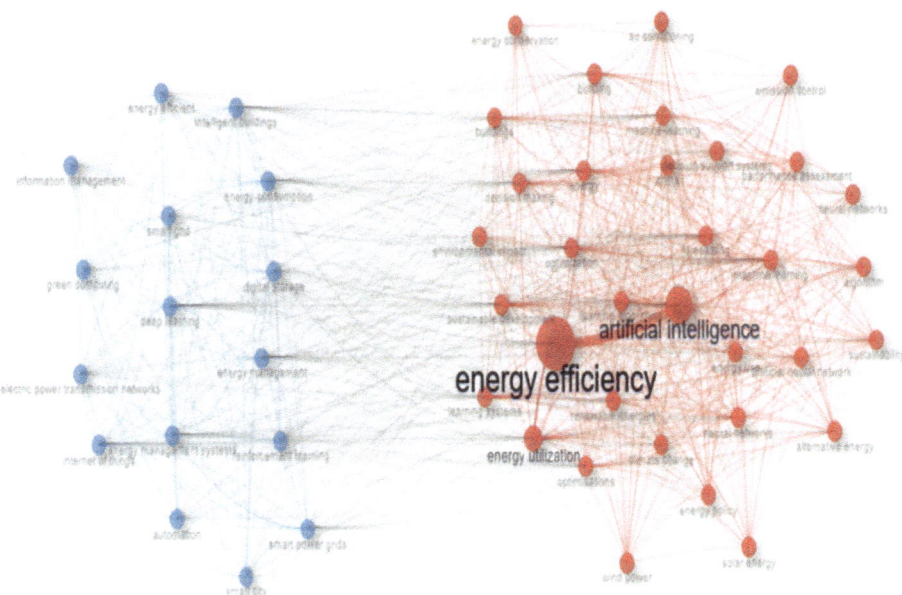

Fig. 2. Keyword co-occurrence network showing thematic clusters in AI and building sustainability research (2022–2025).

The analysis of the 26 selected studies revealed distinct patterns in how Artificial Intelligence (AI) technologies are applied to support sustainability in the built environment. The results are discussed in three interrelated subsections. The first outlines the integration of AI across building lifecycle phases, the second categorizes the types of tools encountered in the literature, and the third reflects on the main purposes these tools serve in promoting sustainable outcomes.

3.2   Integration of AI Across the Building Lifecycle

A core finding of this review is that the application of AI in the context of building sustainability is closely aligned with key phases of the building lifecycle. Most studies demonstrate that AI offers added value not as a standalone digital solution, but as a process-oriented enabler embedded within specific lifecycle stages.

In the Design and Planning phase, AI is frequently used for early-stage performance predictions and optimization [1]. Studies report the deployment of machine learning algorithms to estimate future energy consumption based on material choices, building orientation, and architectural layout [2]. Tools like AI-driven generative design platforms and spatial analysis engines support decision-making by simulating multiple scenarios under constrained parameters, thus allowing for more sustainable design configurations [3]. In several reviewed papers, neural networks and support vector machines were applied to predict heating and cooling loads in the conceptual design stage, improving accuracy and allowing for data-driven material selection and layout adjustments before construction [4].

During the Operation and Monitoring phase, the role of AI shifts toward real-time data analytics and system control [5]. AI models trained on historical building performance data are used to detect anomalies in HVAC systems, predict occupancy patterns, and adjust lighting and temperature settings for optimal comfort and energy efficiency [6]. Smart building platforms integrating reinforcement learning and sensor fusion techniques are gaining ground in this phase. In the literature, operational strategies supported by AI were often cited for enabling proactive maintenance, reducing energy waste, and adjusting building systems based on dynamic environmental inputs or user behavior (Tables 2, 3 and 4).

Table 2. AI Applications Across Building Lifecycle Phases

Lifecycle Phase	AI Applications	Insights from Literature
Design and Planning	Energy forecasting, material selection, generative design, spatial optimization	AI used for early-stage simulation and decision support; neural networks improve accuracy
Operation and Monitoring	Real-time control, anomaly detection, occupant behavior prediction	AI enables dynamic building responses, smart control of HVAC/lighting, and preventive actions
Assessment and Optimization	Retrofit planning, scenario comparison, performance evaluation	AI assists in identifying optimal renovation paths based on energy, cost, and emissions
Compliance and Certification Support	Automated reporting, design validation, regulatory alignment	Tools help meet LEED/BREEAM/EPBD targets; an emerging trend in policy compliance

The Assessment and Optimization phase includes studies where AI supports performance evaluation and renovation planning. Several tools in this domain are designed to model retrofit scenarios by combining building simulation outputs with AI-generated predictions on cost, energy use, and potential emissions reduction [7]. This allows stakeholders to compare multiple renovation pathways and select the most cost- and carbon-efficient option. While life cycle assessment (LCA) was referenced in some studies, it was typically linked to broader performance assessments rather than in-depth LCA automation, as this was not a dominant theme in the final selection of studies [8].

In the final phase, Compliance and Certification Support, AI is applied to streamline the generation of sustainability documentation and facilitate decision-making aligned with certification requirements such as LEED, BREEAM, and EPBD-compliant energy performance certificates. The reviewed studies highlight AI's contribution in cross-referencing design or operational data with certification benchmarks, producing automated compliance reports, or recommending design adjustments to meet minimum energy performance criteria. Although less prevalent than the other phases, this area showed promising growth, particularly in studies targeting policy support and smart regulation compliance tools.

3.3 Typologies of AI Tools in Building Sustainability

A second layer of analysis involved classifying AI tools used in the selected studies into four broad categories: general-purpose AI/ML platforms, data analytics platforms, building-specific AI tools, and applied or niche tools tailored to sustainability contexts.

General AI/ML platforms were widely referenced, especially in studies focused on model training and algorithm development [9]. These tools typically serve as a foundation for building custom solutions and include environments such as TensorFlow, PyTorch, and Scikit-learn. Researchers used these platforms to develop tailored models for energy forecasting, optimization, and anomaly detection.

Table 3. Types of AI Tools in Sustainable Building Research

Tool Category	Examples	Primary Use Cases
General AI/ML Platforms	TensorFlow, Scikit-learn, PyTorch	Model development for prediction, classification, control
Data Analytics Platforms	Power BI, KNIME	Workflow-based visualization, exploratory data analysis
Building-Specific AI Tools	Digital twins, BIM-integrated AI modules	Real-time performance feedback, intelligent control
Applied/Niche Tools	Daylighting optimizers, thermal comfort models	Highly specific applications; support comfort and passive strategies

Data analytics platforms like KNIME or Power BI were noted for their low-code or no-code capabilities [10]. These were primarily used in studies that emphasized accessibility and visualization rather than algorithm development. Their workflow-based interfaces made them suitable for stakeholders without advanced programming expertise, providing intuitive means to track energy performance or conduct exploratory data analysis[11]. Building-specific AI tools represent a growing category that includes digital twins, simulation-coupled AI engines, and intelligent building management systems. These tools are often integrated with Building Information Modeling (BIM) platforms or sensor-based monitoring frameworks, offering real-time AI insights within the operational layer of the building [12]. Examples include AI modules embedded within energy management platforms and smart grid-aware control systems [13]. Applied or niche tools emerged in highly focused sustainability areas such as daylighting optimization, indoor environmental quality assessment, or occupancy-driven thermal control [14]. While these tools tend to be narrower in scope, their targeted nature allows for high precision and relevance [15]. For instance, AI-based daylight prediction models were employed to optimize window placement and shading strategies for improved energy use and comfort [16].

3.4 Application Domains and Functional Roles of AI

The final dimension of analysis concerns the specific purposes AI tools serve in building sustainability. These functions fall into five major domains: prediction and modeling, data analysis and decision support, building performance simulation, sustainability assessment, and operational optimization. Prediction and modeling were the most dominant application domains across the reviewed studies. AI was frequently used to forecast energy consumption, predict occupant behavior, or estimate environmental impacts under future use scenarios. Machine learning algorithms such as random forests, gradient boosting machines, and neural networks were prevalent in this space, offering substantial improvements over traditional statistical methods.

Data analysis and decision support was the second most frequently addressed domain. AI-enabled platforms supported multi-criteria decision-making processes by evaluating trade-offs between environmental, economic, and functional parameters [17]. These tools facilitated design-stage assessments where multiple material or layout options were assessed in parallel, providing actionable insights through visual dashboards or ranking systems [18]. Building performance simulation involved the integration of AI with existing simulation engines like EnergyPlus or TRNSYS. These hybrid tools helped reduce computational time, improve result accuracy, or introduce learning-based adaptability into otherwise static simulation environments [19]. In many cases, simulation data served as training inputs for AI models used in performance optimization or control feedback loops [20]. Sustainability assessment, although referenced in fewer studies, featured in applications that involved system-wide sustainability scoring, material impact estimation, or scenario comparison for retrofitting [21]. While full LCA integration was not a major theme, AI was instrumental in supporting simplified assessments or linking design choices with environmental metrics [22]. Finally, operational optimization and fault detection was a strong area of focus in the Operation and Monitoring phase [23]. AI was used to improve HVAC scheduling, detect faults in

Table 4. Functional Roles of AI in Sustainability Applications

Application Domain	AI Functionality	Benefits Identified in Literature
Prediction and Modeling	Energy use, occupant behavior, cost forecasting	Increases accuracy, enables scenario testing
Data Analysis and Decision Support	Multi-criteria assessment, trade-off analysis	Supports designers and planners in sustainable decision-making
Building Performance Simulation	Hybrid AI-simulation environments	Reduces computation time, adds adaptability to simulation tools
Sustainability Assessment	Impact scoring, simplified evaluations	Facilitates comparison of sustainable alternatives
Operational Optimization and Fault Detection	HVAC control, anomaly detection, energy saving strategies	Improves comfort, reduces operational costs and energy consumption

mechanical systems, and adapt energy management strategies in real time [24]. Studies showed that reinforcement learning and predictive control strategies could reduce energy consumption while maintaining or enhancing occupant comfort levels [25].

4   Conclusions

This study examined how Artificial Intelligence (AI) contributes to the sustainability assessment of buildings, using the building lifecycle as a guiding framework. Through a systematic literature review, it was found that AI applications are increasingly aligned with the four key phases: design and planning, operation and monitoring, assessment and optimization, and compliance and certification support. In the design phase, AI supports early decision-making through energy forecasting and spatial optimization. During operation, AI facilitates real-time monitoring, anomaly detection, and adaptive control of systems. For performance assessment, AI enables scenario modeling and supports renovation planning, while in the compliance phase, it assists in aligning building data with certification requirements such as LEED and EPBD. The review also revealed four main types of AI tools—general-purpose machine learning platforms, data analytics environments, building-specific tools, and niche applications—each contributing uniquely across functional domains, including prediction, decision support, simulation, and optimization. Importantly, AI is not a standalone solution but acts as an enabling layer integrated into broader decision and control systems within the built environment. While this review captured the current landscape of AI-driven sustainability assessment, future work should focus on the validation of AI models in real-world settings and their long-term impact on building performance.

References

1. Mahanta, N.R., Lele, S.: Evolving trends of artificial intelligence and robotics in smart city applications: Crafting humane built environment. Trust-Based Communication Systems for Internet of Things Applications pp. 195–241 (2022)
2. Kim, S.H., Joo, H.J., Kim, J.Y., Kim, H.J., Park, E.-C.: Healthcare policy agenda for a sustainable healthcare system in Korea: building consensus using the Delphi method. J. Korean Med. Sci. **37** (2022)
3. Afzal, M., et al.: Delving into the digital twin developments and applications in the construction industry: a PRISMA approach. Sustainability (Switzerland), **15** (2023)
4. Smolansky, A., Cram, A., Raduescu, C., Zeivots, S., Huber, E., Kizilcec, R.F.: Educator and student perspectives on the impact of generative AI on assessments in higher education. L@S 2023 - Proceedings of the 10th ACM Conference on Learning @ Scale, pp. 378–382 (2023)
5. Sleem, M.M., Abdelfattah, O.Y., Abohany, A.A., Sorour, S.E.: A comprehensive approach to biodiesel blend selection using GRA-TOPSIS: a case study of waste cooking oils in Egypt. Sustainability (Switzerland), **16** (2024)
6. Borchers, M., Gierlich-Joas, M., Tavanapour, N., Bittner, E.: Let citizens speak up: Designing intelligent online participation for urban planning. Lecture Notes Comput. Sci. 14621 LNCS, 18–32 (2024)
7. Cerchione, R., Morelli, M., Passaro, R., Quinto, I.: A critical analysis of the integration of life cycle methods and quantitative methods for sustainability assessment. Corp. Soc. Responsib. Environ. Manag. **32**, 1508–1544 (2025)
8. Stephen, S., Aigbavboa, C., Oke, A.: Revolutionising green construction: harnessing zeolite and AI-driven initiatives for net-zero and climate-adaptive buildings. Buildings **15** (2025)
9. Sayed, A., Himeur, Y., Bensaali, F., Amira, A.: Artificial intelligence with IoT for energy efficiency in buildings. Emerg. Real-World Appl. Internet Things 233–252 (2022)
10. Syed Ahmad, S.S., Yung, S.M., Kausar, N., Karaca, Y., Pamucar, D., Al Din Ide, N.: Nonlinear integrated fuzzy modeling to predict dynamic occupant environment comfort for optimized sustainability. Sci. Programm. (2022)
11. Bhuvana, J., Iqbal, M.A., Batra, R.: Moving towards sustainable smart cities: a review. 2023 International Conference on Power Energy, Environment and Intelligent Control (PEEIC 2023), pp. 1201–1206 (2023)
12. Gerlach, J., Lier, S.K., Hoppe, P., Breitner, M.H.: Critical success factors for AI-driven smart energy services. 29th Annual Americas Conference on Information Systems (AMCIS 2023) (2023)
13. Alhashimi, R.: Artificial intelligence and architecture: Exploring the intersection. Lecture Notes Netw. Syst. **1080 LNNS**, 397–401 (2024)
14. Sharma, A., Kulshrestha, P.: An AI and IoT based smart green home sustainability. Lecture Notes Netw. Syst. **945** LNNS, 59–70 (2024)
15. Gachkar, D., Gachkar, S., García Martínez, A., Angulo, C., Aghlmand, S., Ahmadi, J.: Artificial intelligence in building life cycle assessment. Archit. Sci. Rev. **67**, 484–502 (2024)
16. Yang, F., Chang, H.: Understanding the seesaw effects: pollutant substitution in the process of energy consumption, water pollution, and welfare maximization. Process Saf. Environ. Prot. **194**, 1221–1234 (2025)
17. Vegter, M.W., Blok, V., Wesselink, R.: Process industry disrupted: AI and the need for human orchestration. J. Responsible Technol. **21** (2025)
18. Stojanovski, T., et al.: Rethinking computer-aided architectural design (CAAD)–From generative algorithms and architectural intelligence to environmental design and ambient intelligence. Commun. Comput. Inf. Sci. **1465**(CCIS), 62–83 (2022)

19. Strengers, Y.: AI at home: An urgent urban policy and research agenda. Urban Policy Res. **40**, 250–258 (2022)
20. Donati, F., et al.: The future of artificial intelligence in the context of industrial ecology. J. Ind. Ecol. **26**, 1175–1181 (2022)
21. Markert, J., Saubke, D., Krenz, P., Hotz, L.: Cross-company routing planning: determining value chains in a dynamic production network through a decentralized approach. Proceedings of the Conference on Production Systems and Logistics, pp. 277–286 (2022)
22. Ansari, F., Kohl, L.: AI-enhanced maintenance for building resilience and viability in supply chains. Springer Ser. Supply Chain Manage. **20**, 163–185 (2022)
23. Chen, Z., Zhou, Y., Huang, Z., Xia, X.: Towards efficient multiobjective hyperparameter optimization: a multiobjective multi-fidelity Bayesian optimization and Hyperband algorithm. Lecture Notes Comput. Sci. **13398**(LNCS), 160–174 (2022)
24. Cédric Cabral, F.Y., Patrick Joël, M.M., Ursula Joyce Merveilles, P.N., Marcelline Blanche, M., Georges Edouard, K., Chrispin, P.: Using AI as a support tool for bridging construction informal sector mechanisms to sustainable development requirements. J. Decis. Syst. **31**, 226–240 (2022)
25. Murthy Nimmagadda, S., Harish, K.S.: Review paper on technology adoption and sustainability in India towards smart cities. Multimedia Tools Appl. **81**, 27217–27245 (2022)

Urban Digital Twin Data Requirements and Reference Architecture for Green Spaces and Ecosystems

Lina Morkunaite[1]([✉]), Darius Pupeikis[1], Vytautas Bocullo[1], Egle Klumbyte[1], Andrea Conserva[2], Chiara Farinea[2], Alice Bazzica[2], Peter Barmann[3], and Fruzsina Csala[2]

[1] Faculty of Civil Engineering and Architecture, Kaunas University of Technology, Kaunas, Lithuania
`lina.morkunaite@ktu.lt`
[2] Institute for Advanced Architecture Catalonia, Barcelona, Spain
[3] Sensative AB, Lund, Sweden

Abstract. As cities face growing pressure from climate change, biodiversity loss, and urbanization, there is an urgent need for data-driven tools to support the planning and resilience of green infrastructure. This paper presents a methodology for developing Urban Digital Twins (UDTs) focused on green space planning, monitoring, and regeneration. The process begins with the identification of Key Performance Indicators (KPIs) across five thematic areas: pollution and climate, natural environment, ecosystems, human perception, and public awareness. Based on these KPIs, a set of enabling digital technologies is evaluated through expert ranking, using Kendall's W concordance coefficient to assess consensus on their relevance. The results highlight strong agreement among experts, with IoT, GIS, and BIM emerging as the most suitable technologies due to their capacity for real-time sensing, semantic integration, and spatial representation. Drawing from these insights, the paper proposes a five-layer reference architecture for UDTs designed to support adaptive, inclusive, and data-driven urban greening efforts. The findings offer guidance for cities and stakeholders aiming to implement UDTs for urban resilience.

Keywords: Digital Twin · Photogrammetry · IoT · Green spaces · Urban Areas · Natural Environments

1 Introduction

Urban Digital Twins (UDTs) are increasingly recognized as transformative tools in urban planning and environmental management, offering dynamic and data-driven insights into the functioning and resilience of complex urban ecosystems [1]. As cities face growing challenges related to climate change, biodiversity loss, pollution, and the degradation of green infrastructure, there is a pressing need for tools that can support integrated, evidence-based decision-making [2].

© The Author(s) 2026
A. Jurelionis et al. (Eds.): BDTIC 2025, LNCE 775, pp. 50–62, 2026.
https://doi.org/10.1007/978-3-032-09040-9_5

Traditional urban monitoring systems often operate in silos, lack interoperability, and fail to account for the interconnected nature of environmental, social, and spatial processes [3]. In this end, UDTs provide a promising solution by enabling the real-time integration of heterogeneous data into a coherent, interactive digital representation of urban environments. This makes them particularly valuable for assessing and enhancing the performance of green spaces and ecosystems in cities [4].

This study, as part of *Horizon Europe GreenInCities* project, adopts UDTs as a core methodological component for fostering urban sustainability and resilience. UDTs play a critical role in all co-creation phases by enabling citizens to interact with virtual representations of the pilot areas and actively participate in decision-making [2]. This approach aims at enhancing transparency and inclusivity and ensuring that urban regeneration efforts are data-driven and responsive to local needs.

This paper presents a structured approach to developing UDTs for green spaces and ecosystems, beginning with the identification of Key Performance Indicators (KPIs) and use cases, which are then used to identify relevant enabling technologies. Expert rankings, analysed using Kendall's W coefficient, are employed to evaluate the suitability of these technologies for each KPI group. Based on these insights, the paper proposes a reference UDT architecture that integrates real-time sensing, semantic and spatial data representation, and participatory features to support collaborative, data-driven urban greening efforts.

2 State-Of-The-Art

2.1 Digital Twin Concept

In recent years there has been a growing attention to digital twins (DT) in various domains and is now recognised as a key enabler for transformation to Industry 4.0 [5]. The term digital twin was introduced by Michael Grieves in 2003 [6] defining a DT as a digital representation of a physical process, person, place, system or device [7]. DTs are increasingly used in public space design to enhance urban planning, infrastructure management, and citizen engagement [8]. Recently DT has become a buzzword in urban studies, appearing alongside concepts such as cloud computing, big data, smart cities, artificial intelligence, and the Internet of Things (IoT) reflecting on the growing need for digital infrastructure in city planning [9].

Only successful data transfer to and from the physical environment enables the creation of an abstract virtual view. Bidirectional data transfer demands expertise in sensor calibration and measurement, along with proficiency in communication protocols to ensure accurate time stamping and evaluation of data latency and temporal validity. Therefore, more advanced information reconstruction processes are needed in the event of update failures or presence of divergent information [10].

2.2 Digital Twins and Smart Cities

Recent studies highlight the importance of DT in urban development, enabling city managers to model and optimise urban environments through predictive analytics, real-time simulations, and data-driven governance strategies [8]. The concept of UDTs has gained significance for its potential to transform how smart city systems are managed and planned [11]. There are many scenarios where DT can useful: it can be used in disaster simulations and disaster prevention [12], model unique landscapes like wetlands [13], or plan eco-friendly urban environments [14].

The strength of UDTs lies in their strategic simplification and functionality. UDT provides valuable insights into the relationships between demographics, spatial distribution, and quality of life. Many case studies show that by constructing a UDT from the ground up, integration of data and analytics can be achieved, offering evidence-based support to urban planners, policymakers and other stakeholders [14].

Wang et. al. Researched UDT application increasing city resilience to natural disasters. According to the authors UDT can help to enhance the resilience of cities and communities against natural disasters [12]. UDT integration within smart cities enables adaptive governance, where cities can monitor urban systems, simulate potential policy impacts, and proactively manage infrastructure [8]. By leveraging IoT networks and Big Data analytics, they allow city managers to test scenarios, predict urban behaviours, and optimize urban services [8].

EU has been at the forefront of integrating DT technology into urban development through key initiatives such as *Horizon Europe* and the *Smart Cities and Communities* initiative. Under *Horizon Europe's* funding program for research and innovation, numerous projects are exploring how DT can enhance urban planning, environmental management, and infrastructure resilience. These projects aim to create more data-driven and adaptive urban environments by leveraging real-time data, AI-driven simulations, and predictive modelling. Although large European cities often have up-to-date digital cartography and georeferenced urban information, this is less common in medium or small cities. Furthermore, open data may not always be updated or made accessible to users. Also, fieldwork remains crucial for collecting valuable data that would otherwise be unavailable [14].

2.3 Digital Twins for Natural Environments

The natural environment is a highly complex system that is under pressure of climate change [15]. Recently, researchers have started creating DTs for natural environments. Virtual models are developed to understand, manage and protect ecosystems which have huge potential in resisting their global decline in the face of the environmental crisis [10]. Now DT are being adapted in planning and expanding green areas, monitoring factors relevant to wellbeing of a local flora [10, 13–15].

There are few cases of DTs application in agriculture [17]. Verdouw et. al. Studied how DTs can advance smart farming. In this study case conceptual framework compromising a control model based on a general systems approach and an implementation model for DT systems based on the Internet of Things—Architecture (IoT-A), a reference architecture for IoT systems have been proposed [18].

All this can be achieved when DT is provided necessary live data through IoT and similar technologies, so essential KPIs could help tracking the condition of the site and provide data for simulations and analysis. Data collection for flora models often relies on remote sensing techniques such as LiDAR or multispectral imaging to capture data on plant structures and growth. On the other hand, monitoring local animals requires different tools i.e. GPS collars for wildlife monitoring or fixed monitoring stations equipped with e.g. camera traps or radiofrequency identification readers. If DT rely on updating data to estimate population sizes or species compositions within defined areas, modelling challenges can arise due to the movement of new organisms into the area after the model has been defined. This issue is particularly pronounced in fauna models, where animal mobility introduces variability in shorter time scales respective to flora [16].

The goal of this research is to identify the most relevant technologies that often used in the DT architecture for the development and monitoring green areas in urban spaces.

3   Methods

3.1   KPIs for UDT

The development of UDTs begins with the identification and selection of use cases, which guide the functional and non-functional requirements and the determination of required data types and enabling technologies. Each use case is linked to a set of KPIs, which serve as measurable targets. The KPIs used in this study are presented in Table 1, which are grouped into five groups relevant to the urban environment. More detailed descriptions and UDT use cases of KPIs groups are available in Appendices.

Table 1. UDT for green spaces KPIs.

KPIs group	KPIs
Pollution, climate	Air Quality; Light Pollution; Noise; Air Temperature; Wind; Humidity (air/soil).
Environment	Natural Cooling performed by vegetation in urban spaces; Microclimatic Performance; Amount of water used; Permeable Surface; Canopy cover by vegetation; Carbon sequestration by vegetation; Oxygen production by vegetation; Pollutants removal by vegetation; Ozone removal by vegetation; Avoided Runoff by vegetation; Perceived air quality and temperature.
Ecosystems (non-human)	Habitat typologies and extension; Fauna and Flora mapping; Number of active nests; Health of plants and crops; Ecological connectivity.
Perception (human)	Perceived space safety; Flows of people; Accessibility; Safety; Vitality; Equipment (benches etc.); Care; Gender representation; Place functions and activities; Storytelling and Narratives; Subjective perception; Sentiments; Comfort in the space.
Awareness (human)	Citizens involved in educational activities about climate change; co-analysis and co-design activities; in citizens science and co-monitoring; in green practices.

3.2 Digitalization Technologies

UDT can be characterized as a combination of digitization technologies that serve relevant use cases. This study is based on the assessment of popular built environment digitization technologies, which are presented in Table 2, describing their representational purpose, data formats, and other important characteristics.

Table 2. UDT digitalization technologies.

Technology	Description	Representation	Data type, update frequency, format
Photogrammetry	3D spatial reconstruction from images	Geometric and visual representation	Static 3D mesh. Formats: fbx, dae, glTB, kml, i3s.
Spatial thermography scanning	Photogrammetry from IR (infrared thermal images)	Geometric, visual and semantic representation	Static 3D mesh. Formats: fbx, dae, glTB, kml, i3s.
LiDAR	Light Detection and Ranging. Laser scanning.	Geometric representation	Static/dynamic 3D point cloud. Formats: las, laz, ply, e57, csv, txt, rcp.
360 RGB photography	Panoramic image capture	Visual representation	Static/dynamic 360 photos and videos. Formats: jpg, png, tiff, vrphoto, mp4.
RGB photography	Image capture	Visual representation	Static/dynamic photos and videos, orthophoto. Formats: jpg, bmp, png, tiff, avi, mp4.
IRT	Infrared (IR) thermography	Visual and semantic representation	Static/dynamic IR photos, IR videos. Formats: jpg, tiff, png, bmp, avi, mp4.
BIM	Building Information Modelling	Geometric, semantic, and visual representation	Static, standardized 3D building elements with attributes. Formats: ifc, rvt, dgn, pln, skp, nwd, cube.
GIS	Geographic Information System	Geometric and semantic representation.	Static, standardised 2D/3D e-ements with attributes. Formats: shp, GeoJSON, kml, cityGML.

(continued)

Table 2. (*continued*)

Technology	Description	Representation	Data type, update frequency, format
IoT	Internet of Things	Semantic representation	Dynamic, near real-time data, API and historical time-series data. Formats: JSON, xml, csv, xlsx, sql.
CAD	Computer Aided Design	Geometric representation	Static 2D drawings, schemes, plans, etc. Data formats: dwg, dxf, stp, stl.

3.3 Expert Evaluation Methodology for UDT Technologies Ranking

To prioritise digitalisation technologies across different sustainability-related urban use cases, an expert-based ranking method was applied. The approach aimed to identify which digital tools are perceived as most relevant by experts while developing UDT for green spaces in relation to climate, environment, biodiversity, perception, and awareness use case groups.

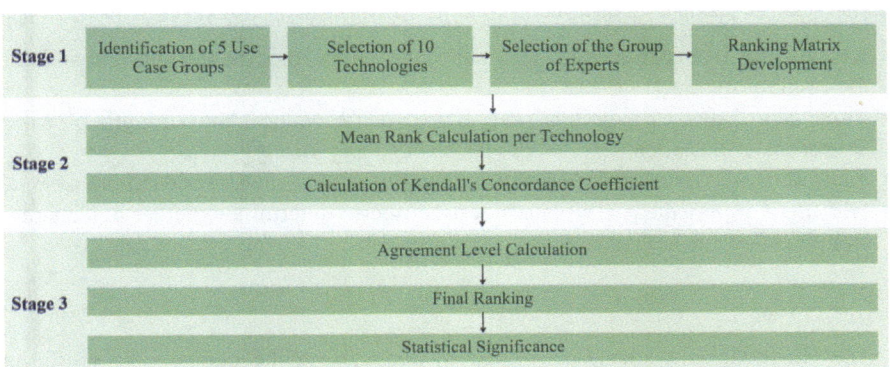

Fig. 1. Expert evaluation methodology for ranking digitalization technologies.

Figure 1 illustrates a three-stage workflow for identifying and ranking digital technologies based on expert evaluations across urban sustainability use case groups.

1) **Setup and Data Collection.** This stage involves preparatory steps necessary for expert elicitation and data gathering:
   - *Identification of Use Case Groups.* In this study, 5 groups were selected (Pollution, Environment, Ecosystems, Perception, Awareness). Each group represents a distinct domain of urban sustainability.

- *Selection of 10 Technologies.* Technologies like IoT, GIS, BIM, Photogrammetry, etc., are predefined as candidates for evaluation.
- *Selection of the Group of Experts.* A panel of 8 qualified experts with experience in smart cities, environmental monitoring, BIM, and geospatial technologies participated in the evaluation. Each expert independently ranked 10 digitalisation technologies from most (rank 1) to least (rank 10) suitable for each of the defined use case groups.
- *Ranking Matrix Development.* Each expert ranks all technologies from 1 (least suitable) to 10 (most suitable) for each use case group. These ranks are compiled into matrices for statistical processing.

2) **Analytical Evaluation.** This stage translates qualitative expert input into structured, quantitative indicators:

- *Mean Rank Calculation per Technology.* The average rank score is computed across all experts for each technology.
- *Calculation of Kendall's Concordance Coefficient (W).* Measures the level of agreement among experts. Values close to 1 indicate strong consensus; values near 0 imply low agreement. The value of the concordance coefficient W is calculated according to the formula [19]:

$$W = \frac{12S}{r^2(n^3 - n)} \tag{1}$$

In the above formula: S is the sum of the squares of the deviations of the sum of the ranks of the performance criteria from the overall mean of the ranks, r is the number of experts, n is the number of criteria.

3) **Results and Interpretation.** This final stage summarizes the insights and validates their reliability:

- *Agreement Level Calculation* [19]. Based on W values, the strength of expert consensus is classified (e.g., moderate, strong, very strong).
- *Final Ranking.* Technologies are ranked by their mean scores for each use case.
- *Statistical Significance.* Chi-square test is used to determine whether agreement is significant or due to chance.

$$X^2 = N(m - 1)W \tag{2}$$

The chi-square statistic is calculated by multiplying the number of experts, the reduced number of objects, and the value of Kendall's coefficient. The resulting X^2 value is then compared to the critical chi-square value with $m - 1$ degrees of freedom. In this way, the null hypothesis H_0, which states that the experts' rankings are random, is tested. If the calculated X^2 exceeds the critical value (at a selected significance level, e.g., $\alpha = 0.05$), the null hypothesis is rejected, leading to the conclusion that the experts' evaluations are statistically significantly aligned [19].

4   Results and Discussion

4.1   Technologies Evaluation Model

The expert-based ranking revealed clear preferences among digitalisation technologies across five sustainability-related urban use case groups. Each group was assessed independently by eight experts, and their rankings were analysed using Kendall's W concordance coefficient to determine the level of agreement (see Appendices). The strength of consensus varied slightly across use case groups, with Kendall's W values ranging from 0.720 to 0.848. All values exceeded 0.7, indicating a strong or very strong level of agreement among experts (Table 3).

Table 3. Expert Consensus on Preferred Technologies by Use Case Group (Kendall's W Coefficient and Agreement Level).

Use Case Group	Kendall's W	Agreement Level	Most Preferred Technologies (Top 3)
Pollution & Climate	0.771	Strong	IoT, GIS, BIM
Environment	0.720	Moderate–Strong	IoT, GIS, Photogrammetry
Ecosystems (non-human)	0.779	Strong	Photogrammetry, GIS, IRT
Perception (human)	0.793	Strong	IoT, GIS, BIM
Awareness (human)	0.848	Very Strong	IoT, GIS, BIM

Across all groups, several technologies consistently appeared among the top preferences. Internet of Things (IoT) and Geographic Information Systems (GIS) were ranked in the top 3 in all five use case groups, highlighting their perceived versatility and applicability across diverse urban sustainability contexts. Building Information Modelling (BIM) also emerged as a strong performer, especially in human-focused domains such as Perception and Awareness. Photogrammetry showed strength in environmental and ecosystem-related groups, reflecting its value in spatial analysis and biodiversity monitoring. These results demonstrate a clear expert consensus regarding the digital technologies best suited to support various sustainability goals in urban contexts.

4.2   Urban Digital Twin Architecture

Based on the proposed KPIs for urban green spaces, resilience, ecosystems, and public spaces—alongside the enabling technologies identified in Sect. 3.2—a reference architecture is introduced (Fig. 2). The architecture comprises five main layers: Data Acquisition, Storage, Data Processing & Integration, Simulation & Analytics, and Visualization, interconnected via standardized transmission protocols, ETL (Extract, Transfer, Load) processes, and APIs. A cross-cutting Security Layer ensures system-wide integrity and trust.

The Data Acquisition Layer collects multi-source data across three domains: (1) processed 2D/3D visual and geometric data with semantic annotations (e.g., photogrammetry models, BIM), (2) real-time environmental and operational data from IoT sensors, and (3) auxiliary sources such as user input, access logs, and weather data. The Storage Layer adopts a multi-database approach, based on recent scientific contributions [20–22]. Object databases handle unstructured assets (e.g., images, meshes); graph databases manage ontologies and semantic links; time-series databases store high-frequency sensor streams; and NoSQL systems support flexible data types. A central Catalogue Graph, as proposed in [23], acts as a semantic registry aligned with Linked Open Data principles.

The Data Processing & Integration Layer ensures harmonization, enrichment, and semantic structuring of incoming data through ETL pipelines, validation, and meta-data enhancement. Domain-specific knowledge graphs and standard ontologies enable semantic reasoning, while interoperability is supported through APIs. The Simulation & Analytics Layer includes modular tools aligned with the selected KPIs: pollution and climate models, microclimate and resilience simulations, BIM-based energy analysis, and biodiversity assessments. Additional components cover mobility, accessibility, and behaviour modelling. AI/ML (artificial intelligence/ machine learning) modules support forecasting of trends such as urban heat island effects (UHI), pollution, or pedestrian dynamics.

The Visualization Layer supports multi-modal interaction, including real-time dash-boards, 3D urban models, semantic query interfaces, 2D GIS maps, and immer-sive AR/VR (augmented/virtual reality) tools for public engagement and co-design. Finally, the Security Layer ensures data confidentiality, integrity, and controlled access through encryption, token-based authentication, and role-based access control, securing interactions across the entire architecture.

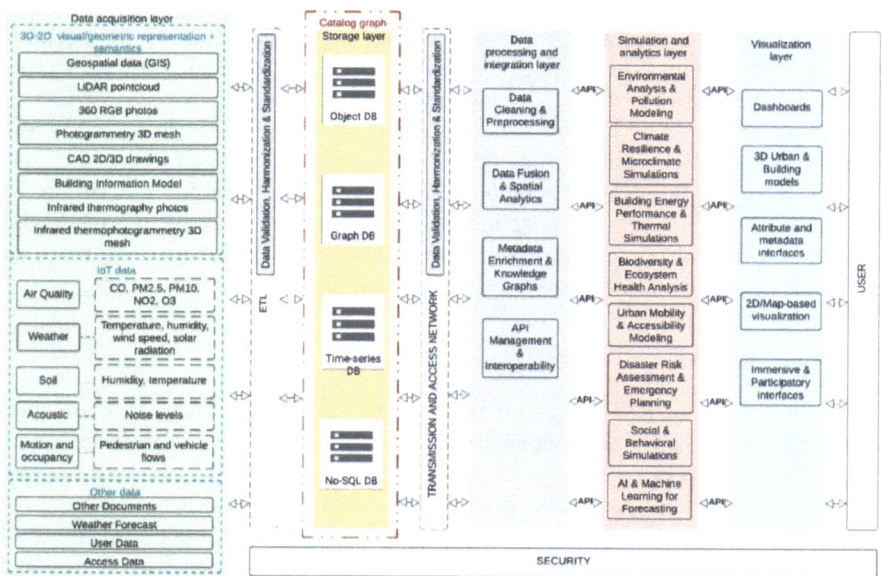

Fig. 2. Urban Digital Twin architecture.

4.3   Contribution of UDT in Enhancing Participatory Processes

Despite their technological complexity, UDTs offer intuitive and accessible tools that can democratize urban decision-making. By providing interactive, data-driven models, UDTs support participatory urban regeneration across all phases of co-creation process, enhancing communication, community involvement, and public trust. Through platforms such as web interfaces, virtual and augmented reality, diverse stakeholders—including residents, planners, and policymakers—can engage directly with urban projects.

UDTs improve the visualization of urban plans by offering interactive 3D models that make regeneration scenarios easier to understand and comment on, supporting informed feedback (Co-analysis and Co-design). They enable real-time monitoring of environmental, ecological, and social indicators via IoT data, allowing citizens and planners to assess ongoing changes and outcomes (Co-analysis and Co-monitoring). Gamification features such as voting tools and interactive simulations encourage creative citizen engagement in design processes (Co-design). During implementation, UDTs increase transparency by visualizing construction progress and site updates in real time (Co-implementation). Post-project, they facilitate collaborative monitoring and maintenance, enabling communities to report issues, propose uses for public spaces, and track ongoing improvements (Co-maintenance and Co-monitoring). Furthermore, UDTs make urban planning more inclusive by offering remote access to information, removing barriers for those who cannot participate in person (All phases). By embedding UDTs into participatory processes, cities can foster more inclusive, transparent, and responsive regeneration strategies that better reflect citizen needs and strengthen connections between communities and their evolving urban environments.

5   Conclusions

This study demonstrates the value of expert-based evaluation as an approach to prioritizing digital technologies for UDT applications in green space and ecosystem-focused urban planning. By engaging eight experts and applying Kendall's W coefficient to assess consensus, the study achieved results with concordance values ranging from 0.720 to 0.848 across five thematic KPI groups. These values indicate strong to very strong agreement and confirm that the rankings are both consistent and meaningful. The evaluation revealed a clear expert consensus around the relevance of IoT, GIS, and BIM technologies. These were consistently ranked among the top tools for supporting data acquisition, integration, and visualisation in UDTs.

By grounding technology selection in expert judgment, the study offers not only a validated prioritization of tools, but also a decision-support mechanism for UDT planning in cities. The expert evaluation process is further supported by a modular five-layer UDT architecture. This architecture complements the prioritized technologies by providing a foundation for implementing responsive, participatory, and data-informed urban regeneration.

6   Limitations and Future Work

This study provides a conceptual framework for developing UDTs for green spaces and ecosystems, but several limitations should be noted. The expert-based ranking, while supported by Kendall's W analysis, remains subjective and may benefit from a broader and more diverse panel. The proposed architecture has not been validated through real-world implementation, leaving practical integration, performance, and user interaction untested. While key enabling technologies were considered, emerging tools such as edge computing or advanced AI were beyond the scope. Lastly, social and institutional factors like data governance and digital equity warrant further exploration in future work.

7   Appendices

The expert evaluation materials—including KPI descriptions, use case context, expert ranking sheets, and the corresponding Kendall's W concordance coefficient calculations—are available via the Mendeley Data repository under the following https://doi.org/10.17632/9whh6dy5j9.2.

Acknowledgements. This study is part of the dissemination activities of the research projects "Demonstrating Holistic Data-driven Co-Creative Approaches in Nature-Based Solutions towards Climate Adaptation and Mitigation (GreenIn Cities)" (Grant ID Number 101139730), funded by Horizon Europe call HORIZON-MISS-2023-CLIMA-CITIES-01.

References

1. D. Gürdür Broo, M.B.-H., Schooling, J.: Design and implementation of a smart infrastructure digital twin. Autom. Constr. **136**(January), 104171 (2022). https://doi.org/10.1016/j.autcon.2022.104171
2. Kahlen, F.-J., Flumenrfelt, S., Alves, A.: Transdisciplinary perspectives on complex systems. Springer (2016)
3. White, G., Zink, A., Codecá, L., Clarke, S.: A digital twin smart city for citizen feedback. Cities **110**(January) (2021). https://doi.org/10.1016/j.cities.2020.103064
4. Mazzetto, S.: A review of urban digital twins integration, challenges, and future directions in smart city development. Sustain. **16**(19), (2024). https://doi.org/10.3390/su16198337
5. Batty, M.: Digital twins. Environ. Plan. B Urban Anal. City Sci. **45**(5), 817–820 (2018). https://doi.org/10.1177/2399808318796416
6. Kritzinger, W., Karner, M., Traar, G., Henjes, J., Sihn, W.: Digital Twin in manufacturing: a categorical literature review and classification. IFAC-PapersOnLine **51**(11), 1016–1022 (2018). https://doi.org/10.1016/j.ifacol.2018.08.474
7. Maimour, M., Ahmed, A., Rondeau, E.: Survey on digital twins for natural environments: a communication network perspective. Internet Things (Netherlands) **25**(July 2023), 101070 (2024). https://doi.org/10.1016/j.iot.2024.101070
8. Weil, C., Bibri, S.E., Longchamp, R., Golay, F., Alahi, A.: Urban digital twin challenges: a systematic review and perspectives for sustainable smart cities. Sustain. Cities Soc. **99**(July), (2023). https://doi.org/10.1016/j.scs.2023.104862

9. Wang, Y., et al.: Digital twin approach for enhancing urban resilience: a cycle between virtual space and the real world. Resilient Cities Struct. **3**(2), 34–45 (2024). https://doi.org/10.1016/j.rcns.2024.06.002

10. Chen, H., et al.: Digital twin-based virtual modeling of the Poyang Lake wetland landscapes. Environ. Model. Softw. **181**, (2024). https://doi.org/10.1016/j.envsoft.2024.106168

11. Villanueva-Merino, A., Urra-Uriarte, S., Izkara, J.L., Campos-Cordobes, S., Aranguren, A., Molina-Costa, P.: Leveraging local digital twins for planning age-friendly urban environments. Cities **155**, (2024). https://doi.org/10.1016/j.cities.2024.105458

12. Blair, G.S.: Digital twins of the natural environment. Patterns **2**(10), 100359 (2021). https://doi.org/10.1016/j.patter.2021.100359

13. Paolo, G., et al.: Digital twin enhanced with Machine Learning Algorithms for Irrigation Management Using Sensor Data. In: 6th International Conference on Industry 4.0 and Smart Manufacturing, vol. 253, no. 2024, pp. 2419–2427 (2025). https://doi.org/10.1016/j.procs.2025.01.302

14. Kim, Y., Oh, J., Bartos, M.: Stormwater digital twin with online quality control detects urban flood hazards under uncertainty. Sustain. Cities Soc. **118**, 105982 (2025). https://doi.org/10.1016/j.scs.2024.105982

15. Mrosla, L., Fabritius, H., Kupper, K., Dembski, F., Fricker, P.: Review on the dynamic modelling of flora and fauna in digital twins. Ecol. Modell. **504**(November), 2025 (2024). https://doi.org/10.1016/j.ecolmodel.2025.111091

16. Pylianidis, C., Osinga, S., Athanasiadis, I.N.: Introducing digital twins to agriculture. Comput. Electron. Agric. vol. **184**(October 2020), 105942 (2021). https://doi.org/10.1016/j.compag.2020.105942

17. Verdouw, C., Tekinerdogan, B., Beulens, A., Wolfert, S.: Digital twins in smart farming. Agric. Syst. **189**(December 2020), 103046 (2021). https://doi.org/10.1016/j.agsy.2020.103046

18. Schlenger, J., Pluta, K., Mathew, A., Yeung, T., Sacks, R., Borrmann, A.: Reference architecture and ontology framework for digital twin construction. Autom. Constr. **174**, 106111 (2025)

19. Lu, Q., et al.: Developing a digital twin at building and city levels: case study of West Cambridge campus. J. Manag. Eng. **36**(3), 05020004 (2020)

20. Kendall, M.G.: rank correlation methods:Book. London, Griffin, 202. (1970). ISBN 10:085264199

21. Ramonell, C., Chacón, R., Posada, H.: Knowledge graph-based data integration system for digital twins of built assets. Autom. Constr. **156**, 105109 (2023)

22. Chamari, L., Petrova, E., Pauwels, P.: An end-to-end implementation of a service-oriented architecture for data-driven smart buildings. IEEE Access **11**, 117261–117281 (2023)

23. Knezevic, M., Donaubauer, A., Moshrefzadeh, M., Kolbe, T.H.: Managing urban digital twins with an extended catalog service. ISPRS Ann. Photogrammetry Remote Sens. Spatial Inf. Sci. **10**, 119–126 (2022)

Towards True Networked Urban Digital Twins – A Development Agenda

Juho-Pekka Virtanen[1,2(✉)] [iD], Laura Mrosla[3,4,5] [iD], and Tapani Heino[6]

[1] Forum Virium Helsinki, Helsinki, Finland
juho-pekka.virtanen@forumvirium.fi
[2] Finnish Geospatial Research Institute FGI, Helsinki, Finland
[3] FinEst Centre for Smart Cities, Tallinn University of Technology, Tallinn, Estonia
[4] Department of Civil Engineering and Architecture, Tallinn University of Technology, Tallinn, Estonia
[5] Aalto University, Otaniementie 14, 02150 Espoo, Finland
[6] City of Helsinki, Helsinki, Finland

Abstract. Urban Digital Twins (UDT) have become ambitions for many cities globally, yet their implementation varies significantly in terms of governance, openness and their relation with external stakeholders. Currently, many UDTs still rely on manual updating, leading to discussion on whether they really are digital twins by strict definition. To maximize their utility, UDTs must be developed to meet the multiple needs of the cities (e.g. decision making) while ensuring continuous updates and adaptability. In this work, we present a development agenda for UDTs, synthesizing insights from both recent literature and the practical experiences gained through various projects carried out by Forum Virium Helsinki and the FinEst Centre for Smart Cities. In this work, we identified four key elements for the future development of UDTs: 1) establishing common understanding, 2) modularity & interoperability 3) presentation agnosticism and 4) the social dimension.

Keywords: Urban Digital Twin · Digital Twin · Smart City · Urban Data · Data Ecosystem

1 Introduction

Urban Digital Twins (UDTs) have emerged with the promise of becoming transformative tools for urban planning and management. As cities globally adopt them, approaches diverge in use cases, governance, openness and their relation with stakeholders (D'Hauwers et al., 2021). UDTs hold the potential to support overcoming urban challenges and complexity by bridging the siloed structures of cities, which are reflected in the data management, city departments, and involved stakeholders (Weil et al., 2023).

However, many of the current UDT implementations appear as stand-alone applications tackling explicit urban challenges, but are not interoperable with each other. Therefore, attaining the full potential requires an ecosystemic approach that integrates stakeholders, technologies, and data across interconnected systems (Bennett et al., 2023).

© The Author(s) 2026
A. Jurelionis et al. (Eds.): BDTIC 2025, LNCE 775, pp. 63–73, 2026.
https://doi.org/10.1007/978-3-032-09040-9_6

Scholars have emphasized the dual role of cities as both physical and social entities (e.g. Batty, 2024; Bettencourt 2024). For UDTs to surpass the capabilities of existing city models and provide meaningful advantages across the multitude of urban planning and governance tasks, a deeper integration of physical and social systems is essential. This integration necessitates a modular, participatory and challenge-driven approach to their implementation (Nochta et al., 2021). Moreover, unlocking the real potential of UDTs requires a shift of the planning of digital twins from systems built for individual and isolated use cases to networked, ecosystemic frameworks.

By designing digital twin applications with interoperability in mind from the early stages, they can be seamlessly connected to form a broader digital twin ecosystem. This approach enhances their value and usefulness, allowing digital twins to move beyond their current role as isolated tools (Bennett et al., 2023).

In this work, we present a development agenda for UDTs, building on our experiences from the City of Helsinki and FinEst Centre for Smart Cities. We base this development agenda on two assumptions: Firstly, we assume that an ecosystemic approach could effectively support UDTs in navigating the dynamic and heterogeneous technological landscape of the cities. Secondly, from the technical perspective, we argue that inter-operability of data and tools will be a central enabling factor in accomplishing this. We firstly review the current state of UDTs, focusing on Helsinki and FinEst Centre for Smart Cities (Sect. 2.), after which we present the key aspects for further development of the UDTs, consisting of establishing common understanding (Sect. 3.1), modularity & interoperability (Sect. 3.2), presentation agnosticism (Sect. 3.3) and the social dimension (Sect. 3.4).

2   Current State of Urban Digital Twins

Since the emergence of the term "Digital Twin", it has been widely adopted in various disciplines, with diverging definitions ranging from static digital models to autonomous cyber-physical systems. The application of digital twins has gained significant traction throughout the last decade in the context of cities and the built environment (Abdelrahman et al., 2025).

For digital twins of cities or urban environments, a variety of terms are used. These include: urban digital twin, digital twin city (Deng et al., 2021), city scale digital twin (Nochta et al., 2021) smart city digital twin (Francisco et al., 2020) and local digital twin (Raes et al., 2025; Villanueva-Merino et al., 2024). Today, there is no consensus on the definition of what exactly an (urban) digital twin is, and many authors have highlighted the plurality of the concept (Abdelrahman et al., 2025; Deren et al., 2021).

Digital Twins of the urban environment, in this publication "Urban Digital Twins" (UDTs), are perceived as virtual replicas of cities that combine data from multiple sources to create a comprehensive digital representation of urban environments. They hold potential to serve as powerful tools for city planners, policymakers, and stakeholders to monitor, simulate, analyze, and optimize urban systems before implementing changes in the real world (Azadi et al., 2025). Specific digital twin applications support urban asset management and participatory practices. These digital representations hold the potential to revolutionize planning and governance by enabling more up-to-date analysis and future projections based on data-driven insights.

Understandably, UDTs have, and are, being researched and implemented globally with varying foci. Stemming from urban planning, UDTs have initially emerged based on static 2D maps and 3D city models. From this static foundation, the digital representations gradually were enhanced with additional information, e.g. semantic information and dynamic data (e.g. Schrotter & Hürzeler, 2020; Lehner & Dorffner, 2020).

While UDTs already integrate previously siloed data within a single application, most remain constrained to specific use-cases, relying on limited input data, models and simulations. This use-case driven approach represents an intermediate stage in the progression towards a comprehensive digital representation of urban environments. Advancing beyond this stage is crucial for developing a holistic representation of urban systems, supporting the development of smarter, more efficient, and sustainable cities.

2.1 Urban Digital Twins in Helsinki

In Helsinki, the development of UDTs has been strongly built upon prior work with 3D city models. Currently, the city maintains two central city model assets: a textured mesh model (updated by a renewed city wide survey) and a CityGML-based model (updated along with the base map). Both of these models cover the entire administrative area and are openly available (Helsinki, 2025). To further integrate these models to various urban planning and management processes, a development project has been launched at the Helsinki Urban Environment Division. The work began with the development of proof-of-concept level prototypes of both tools and processes that utilize the 3D city models and other geospatial data. The completed projects have been documented as videos, with their results guiding further development work in the city administration (Helsinki Urban Environment Division, 2025).

Forum Virium Helsinki, the city's innovation company, has focused on the development of new application areas and data sources of UDTs (Virtanen et al., 2024), further developing the concept of socio-technical UDT (Ruohomäki et al., 2024) and studying the potential for forming a digital twin of urban mobility (Forum Virium, 2024). Currently, several projects are ongoing related to facilitation of data ecosystems (DataLiiKe, 2025), data marketplaces (SEDIMARK, 2025) and data spaces (TFDS, 2025). These activities contribute to the further integration of data ecosystems, data spaces and UDTs.

2.2 Digital Twins at FinEst Centre for Smart Cities

FinEst Centre for Smart Cities is an international and transdisciplinary research and development centre, operating as an independent organization under Tallinn University of Technology (TalTech). Established in 2019 by TalTech, Aalto University, Forum Virium Helsinki and the Estonian Ministry of Economic Affairs and Communications, the centre focuses on enhancing the quality of life in urban areas. Following this aim, the Centre is researching Urban Digital Twins globally and developing (prototype) digital twin applications within local projects and in collaboration with researchers from various international research institutes, citizens, city officials and leaders (Soe, 2017; FinEst Centre Homepage, 2025).

FinEst Centre for Smart Cities has led the GreenTwins project, focussing on two topics: First, the development of prototypes for a dynamic vegetation layers for the

Digital Twins of the cities Tallinn, Estonia and Helsinki, Finland. This layer includes algorithmically generated 3D models of plants, along with data on their growth patterns and seasonal variations under local climatic conditions. Second, GreenTwins brought up two digital twin applications named Virtual Green Planner and Urban Tempo, designed to engage urban stakeholders, especially citizens, in the design of urban green areas (Prilenska et al., 2023; FinEst Centre, 2023).

Research beyond the integration of biotic layers to UDTs, FinEst Centre for Smart Cities led the development of the Renovation Strategy Tool (ReSTO). ReSTO is a digital-twin based platform for municipal decision-makers enabling the optimization of required investments into their building stock, by assessing alternative scenarios across various aspects, including technology, building and infrastructure design, urban planning, and costs. By leveraging digital twin data from the Estonian Building Registry (https://liv ekluster.ehr.ee/ui/ehr/v1) and other public databases, ReSTO allows to determine economically optimal renovation strategies while considering predefined constraints such as budget availability, energy performance targets, and environmental objectives (Arumägi et al., 2023; FinEst Centre, 2025).

In addition to leading its own research projects, the FinEst Centre for Smart Cities, actively participates in multiple international projects developing UDTs, such as the urbanLIFEcircles project. Here, the Centre contributes to the development of digital twin applications for urban biodiversity monitoring, management and modelling (LIFE Public Database, 2025).

3   Key Elements of Future Urban Digital Twins

3.1   Establish Common Understanding and Standards

As already mentioned (Sect. 2.), a concise and universally applied definition of an UDT does not exist yet. Currently, the inconsistent use of the term "urban digital twin" for various differing concepts hampers the advancement of both the discourse and development in the field. This lack of conceptual clarity also risks reducing "digital twin" to another short-living buzzword, rather than a meaningful enduring framework for urban innovation.

Thus, the establishment of a common understanding, clear definition and common standards for UDTs is necessary, to support both discussion and the development itself. In addition, the UDTs should be conceptually positioned in respect to a number of surrounding concepts, such as:

- 3D City Models (Biljecki et al., 2015)
- Urban Dashboards (Kitchin et al., 2016)
- Urban Data Platforms (Soe et al., 2022)
- Metaverse/Cityverse (Kshetri et al., 2024)

The question on which different authors tend to have diverging opinions is whether a digital twin necessitates automatic feedback into the real world or not. While this requirement originates from a fairly agreed-upon definition of a digital twin in manufacturing technology, it has proven to be less applicable to UDTs. It has been noted that many of the UDTs don't meet the criteria of a digital twin, if a strict definition is applied (Metcalfe et al., 2024).

For our purposes, the inclusion of automated actuators bridging the virtual and the digital world are seen as one additional component in the digital twin framework. However, as the development of UDTs is continuously evolving, this distinction may become increasingly relevant in defining technology readiness and technology maturity models in the future.

Standards and information models related to UDTs can be approached from a holistic perspective, aiming for a model covering all of the object types with their semantics, or with a minimalist approach, ensuring only the object identifiers (Ellul et al., 2024). Multiple ongoing attempts to unify the definition of UDTs can be identified, including at least the following ones (Table 1.). While reaching a cross-sectorally accepted definition might not prove to be feasible now, having an internally coherent understanding of what is meant by a UDT would greatly benefit cities by reducing confusion and supporting internal collaboration.

Table 1. Examples of UDT definition work.

Title	Description	URL
DIN SPEC 91607 Digitale Zwillinge	A German community standard defining the urban digital twins for European communities, along with their structure, use and connections.	https://dx.doi.org/https://doi.org/10.31030/3575521
The Roadmap to the Information Management Framework for the Built Environment	From the already concluded Centre for Digital Built Britain, a roadmap for realization of a national digital twin.	https://doi.org/https://doi.org/10.17863/CAM.38227
OGC UDT DWG	A position paper aiming to clarify the concept of Urban Digital Twins and their relation to surrounding concepts.	http://www.opengis.net/doc/dp/UDT
EU Local Digital Twin Toolbox	Public report on the roadmap proposal for the deployment of the LDT toolbox	https://op.europa.eu/s/z4Hf

3.2 Modularity and Interoperability

An ecosystemic UDT leverages the principles of interoperability and modularity to create a collaborative environment where various stakeholders can contribute and benefit from the system (Bennett et al., 2023). This requires at least a certain degree of common understanding (Sect. 3.1). The approach aligns with the concept of data spaces, which provide a shared digital infrastructure for data exchange and collaboration (Gil et al., 2024). Thus, realizing functional data ecosystems is a relevant task for UDT development as well.

Without a true ecosystem of data, applications and tools, UDTs risk remaining as cities' internal urban planning tools. As the problems encountered in urban settings do not follow administrative boundaries, interoperability of tools is also needed to allow collaboration between administrative units and beyond.

To ensure data protection, privacy and compliance with relevant regulations (such as GDPR), robust security measures and ensuring data sovereignty is crucial for maintaining trust among stakeholders and protecting sensitive urban data.

Realizing UDTs as a networked, modular system (Fig. 1.) involving multiple stakeholders requires the adoption of standardized data formats: Adopting common data standards and formats ensures seamless integration and interoperability between diverse data sources and different components of the UDT. Thus, modularity and interoperability are connected. Additionally, an API-driven architecture enables maintaining a clear master data repository and provision of updates by the responsible entity/organisation. Implementing a robust API layer allows for seamless communication between different modules and external systems, and integration of real-time data and new components such as AI agents or simulation engines. This way, different UDTs may also have their respective use cases or foci.

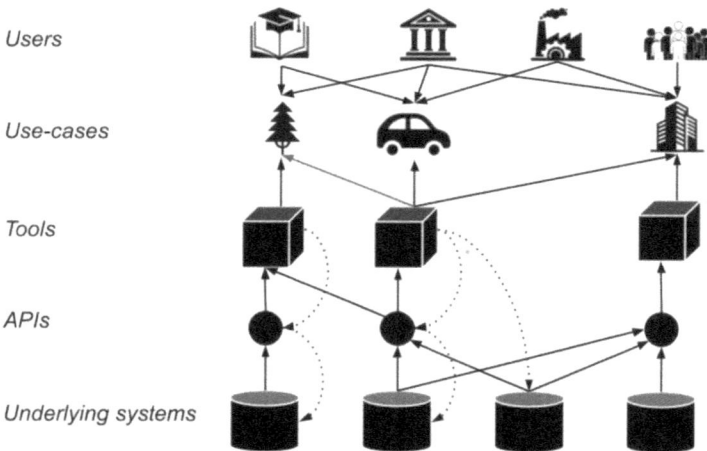

Fig. 1. UDTs formed as an ecosystem of data, APIs and tools enabling data flow across different use cases and users, with returning data fed back to underlying systems.

3.3 Presentation Agnosticism

Digital Twins are primarily presented (and viewed) through various visualizations, including interactive 3D environments, renderings of 3D models and VR/AR. However, their presentation may also incorporate audio feedback, data-driven analytics, and interactive simulations to provide comprehensive and dynamic representations of physical assets or systems (Mrosla et al., 2025).

As UDTs are expected to support decision-making processes, it is essential that they remain agnostic to their means of presentation. This means that the underlying data and models should be independent of any specific presentation tool or technique.

By separating the data and analysis layers from the presentation layer, UDTs can:

1. Accommodate diverse user needs and preferences for (data) presentation.
2. Enable the integration of new presentation technologies as they emerge.
3. Support multiple simultaneous presentations of the same data for different purposes or stakeholders.
4. Support the integration of different external data sources in the presentation medium suited for them.

While the data models commonly used in UDTs, such as CityGML for 3D city models, support both 3D visualization and other types of thematic visualizations (e.g. semantic or analytical representations; Fig. 2.), the implementations of other than geospatial visualization are extremely rare in the UDT context.

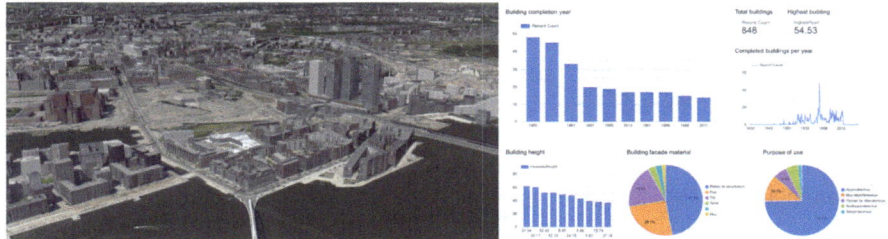

Fig. 2. An example of CityGML building models being visualized in 3D (left) and as a statistical plot of their properties (right), illustrating how same data can be visualized in different tools to serve different use-cases. Data and 3D visualization courtesy of the City of Helsinki.

3.4 Social Dimension

As UDTs are in the end intended to serve citizens, even if by non-direct influence e.g. by providing better decision making environments for the administration, it is worth questioning how well the UDTs are able to cover the complexities of life in cities. Multiple authors argue that current digital twins are merely abstractions of reality, omitting many elements (Batty, 2018). Furthermore, while digital twins are often developed with the intention to enhance public participation e.g. in decision-making processes, this promise remains largely unfulfilled (Charitonidou, 2022).

To address this gap, the "social dimension" in UDTs development and application should be further strengthened, bridging the gap between social processes, social phenomena, urban population and the urban digital twin (Ruohomäki et al., 2024). Following Ruohomäki et al (2024), the social dimension of the UDTs is here understood to include:

- Socio-economic and demographic data
- Inclusion of non-physical spatial artifacts such as legal and administrative boundaries, and management units
- Data from participatory actions and crowdsourcing
- Inclusion of social phenomena and artifacts

The importance of the human dimension is further highlighted by most of the envisioned UDT applications being "human in the loop" systems, supporting decision making in urban planning, management and other operations with improved insights, simulation results and tools (Schrotter & Hürzeler, 2020; Lehner & Dorffner, 2020). While more autonomous systems have been explored in the smart city context, they are typically present in cases related to interpreting sensor data streams from smart infrastructure (Mohammadi & Al-Fuqaha 2018). For processes like urban planning, little is known about how to practically include ethics and other non-tangible values into the potentially autonomous decision making systems–at the same time the role of social processes and artifacts has to be acknowledged for these systems to be truly feasible (Charitodinou, 2022).

4   Discussion and Conclusions

The conception and development of Urban Digital Twins is happening simultaneously at many fronts:

- The technologies involved in digital twins evolve. These include sensor/IoT systems, algorithms, models (and associated standards) and presentation methods such as through VR/AR. Thus, continuous adaptation is needed for the UDTs to leverage the most suitable available and emerging tools.
- Urban planning, management and participatory processes continue to integrate upcoming technologies and data, leading to new and improved use-cases for the UDTs.
- The conceptual discourse on UDTs progresses, potentially leading to a more consensual understanding of what the UDTs actually are.

For this evolving system to be functional, usable and effective, the development of digital twins of cities should not occur haphazardly, depending on available funding and driven by hype-cycles. Instead, proceeding in a well-orchestrated approach, where the intertwined components and their relationships to each other are well thought out, allows forming a functioning, updatable and useful ecosystem of tools, data and technology. Effectively leveraging this in practice can lead to a comprehensive change of how cities are planned and managed: from siloed domain-specific approaches to a holistic understanding of a city.

In this work, we have identified four key elements for future development of UDTs, them being 1) common understanding, 2) modularity & interoperability, 3) presentation agnosticism and 4) the social dimension. While these aims may never be fully agreed upon or reached by all UDT stakeholders, we argue that even a limited progress towards them would be highly beneficial for cities.

Acknowledgements. The work of L.M. has been supported by the Estonian Ministry of Research and Education and European Regional Development Fund (grant 20142020.4.01.20-0289). The work of J.-P.V. has been supported by the European Regional Development Fund, Project "DataLi-iKe", DS4SSCC Project 2024-1-B – TFDS and the Research Council of Finland project no. 359175.

References

Abdelrahman, M., Macatulad, E., Lei, B., Quintana, M., Miller, C., Biljecki, F.: What is a digital twin anyway? Deriving the definition for the built environment from over 15,000 scientific publications. Build. Environ. (2025). https://doi.org/10.1016/j.buildenv.2025.112748

Arumägi, E., Hallik, J., Pikas, E., Kalamees, T., Liiv, I., Kisel, E.: Quantification of building envelope heat losses on a district level for comparative renovation strategies assessment. J. Phys. Conf. Ser. **2654**(1), 012003 (2023). https://doi.org/10.1088/1742-6596/2654/1/012003

Azadi, S., Kasraian, D., Nourian, P., van Wesemael, P.: What have urban digital twins contributed to urban planning and decision making?: A systematic literature review and a socio-technical research and development agenda (2025). https://doi.org/10.3390/smartcities8010032

Batty, M.: Digital twins in city planning. Nat. Comput. Sci. **4**(3), 192–199 (2024). https://doi.org/10.1038/s43588-024-00606-7

Batty, M.: Digital twins. Environ. Plann. B Urban Anal. City Sci. **45**(5), 817–820 (2018). https://doi.org/10.1177/2399808318796416

Bennett, H., Birkin, M., Ding, J., Duncan, A., Engin, Z.: Towards ecosystems of connected digital twins to address global challenges. Zenodo (2023). https://doi.org/10.5281/zenodo.7840266

Bettencourt, L.M.: Recent achievements and conceptual challenges for urban digital twins. Nat. Comput. Sci. **4**(3), 150–153 (2024). https://doi.org/10.1038/s43588-024-00604-9

Biljecki, F., Stoter, J., Ledoux, H., Zlatanova, S., Çöltekin, A.: Applications of 3D city models: state of the art review. ISPRS Int. J. Geo Inf. **4**(4), 2842–2889 (2015). https://doi.org/10.3390/ijgi4042842

Charitonidou, M.: Urban scale digital twins in data-driven society: challenging digital universalism in urban planning decision-making. Int. J. Archit. Comput. **20**(2), 238–253 (2022). https://doi.org/10.1177/14780771211070005

DataLiiKe. https://dataliike.fi/in-english/. last accessed 28 Apr 2025

Deng, T., Zhang, K., Shen, Z.J.M.: A systematic review of a digital twin city: a new pattern of urban governance toward smart cities. J. Manag. Sci. Eng. **6**(2), 125–134 (2021). https://doi.org/10.1016/j.jmse.2021.03.003

Deren, L., Wenbo, Y., Zhenfeng, S.: Smart city based on digital twins. Comput. Urban Sci. **1**, 1–11 (2021). https://doi.org/10.1007/s43762-021-00005-y

D'Hauwers, R., Walravens, N., Ballon, P.: From an inside-in towards an outside-out urban digital twin: business models and implementation challenges. ISPRS Ann. Photogram. Remote Sens. Spatial Inf. Sci. **8**, 25–32 (2021). https://doi.org/10.5194/isprs-annals-VIII-4-W1-2021-25-2021

Ellul, C., Stoter, J., Bucher, B., Olsson, P.O., Billen, R., DeLathouwer, B.: Towards NationalvConnected digital twins: a geospatial perspective. ISPRS Ann. Photogram. Remote Sens. Spatial Inf. Sci. **10**(4/W5-2024), 147–154 (2024). https://doi.org/10.5194/isprs-annals-X4-W5-2024-147-2024

FinEst Centre Homepage. https://finestcentre.eu/. last accessed 28 Mar 2025

FinEst Centre. GreenTwins: Redefining urban greenery design for modern cities (2023). https://finestcentre.eu/media-and-events/news/greentwins-empower-cities-urban-green-areas/. last accessed 28 Mar 2025

FinEst Centre: Renovation Strategy Tool. https://finestcentre.eu/project-pilot/renovation-strategy-tool/. last accessed 28 Mar 2025

Forum Virium: Digital Twin for Mobility - Concept and baseline study (2024). https://forumvirium.fi/wp-content/uploads/2025/01/Digital-Twin-for-Mobilty.-Working-paper-v2-1124.pdf. last accessed 28 Mar 2025

Francisco, A., Mohammadi, N., Taylor, J.E.: Smart city digital twin–enabled energy management: toward real-time urban building energy benchmarking. J. Manag. Eng. **36**(2), 04019045 (2020). https://doi.org/10.1061/(ASCE)ME.1943-5479.0000741

Gil, J., Petrova-Antonova, D., Kemp, G.J.: Redefining urban digital twins for the federated data spaces ecosystem: a perspective. Environ. Plann. B Urban Anal. City Sci. (2024). https://doi.org/10.1177/23998083241302578

Helsinki: Helsinki 3D. https://www.hel.fi/en/decision-making/information-on-helsinki/maps-and-geospatial-data/helsinki-3d. last accessed 28 Mar 2025

Helsinki Urban Environment Division, Kaupunkimalli & digitaaliset kaksoset playlist. https://www.youtube.com/playlist?list=PLN3MiazSNOP3UYAB4uPgGSx5qSKdmlt0d. last accessed 28 Mar 2025

Kitchin, R., Maalsen, S., McArdle, G.: The praxis and politics of building urban dashboards. Geoforum 77, 93–101 (2016). https://doi.org/10.1016/j.geoforum.2016.10.006

Kshetri, N., Dwivedi, Y.K., Janssen, M.: Metaverse for advancing government: prospects, challenges and a research agenda. Gov. Inf. Q. 41(2), 101931 (2024). https://doi.org/10.1016/j.giq.2024.101931

Lehner, H., Dorffner, L.: Digital geoTwin Vienna: towards a digital twin city as geodata hub. PFG 88, 63–75 (2020). https://doi.org/10.1007/s41064-020-00101-4

LIFE Public Database. https://webgate.ec.europa.eu/life/publicWebsite/project/LIFE21-NAT-EE-urbanLIFEcircles-101074453/introducing-adaptive-community-based-biodiversity-man agement-in-urban-areas-for-improved-connectivity-and-ecosystem-health. last accessed 29 Mar 2025

Metcalfe, J., Ellul, C., Morley, J., Stoter, J.: Characterizing the role of geospatial science in digital twins. ISPRS Int. J. Geo Inf. 13(9), 320 (2024). https://doi.org/10.3390/ijgi13090320

Mohammadi, M., Al-Fuqaha, A.: Enabling cognitive smart cities using big data and machine learning: approaches and challenges. IEEE Commun. Mag. 56(2), 94–101 (2018). https://doi.org/10.1109/MCOM.2018.1700298

Mrosla, L., Fabritius, H., Kupper, K., Dembski, F., Fricker, P.: What grows, adapts and lives in the digital sphere? Systematic literature review on the dynamic modelling of flora and fauna in digital twins. Ecol. Model. 504, 111091 (2025). https://doi.org/10.1016/j.ecolmodel.2025.111091

Nochta, T., Wan, L., Schooling, J.M., Parlikad, A.K.: A socio-technical perspective on urban analytics: the case of city-scale digital twins. J. Urban Technol. 28(1–2), 263–287 (2021). https://doi.org/10.1080/10630732.2020.1798177

Prilenska, V., Nummi, P., Tan, X., Mrosla, L., Zarrinkafsh, H., Fabritius, H.: Green-Twins as a Communicative Planning Support System (CPSS). Computational Urban Planning and Urban Management (CUPUM), Montréal, Canada (2023)

Raes, L., Ruston McAleer, S., Croket, I., Kogut, P., Brynskov, M., Lefever, S.: Decide Better: Open and Interoperable Local Digital Twins. (2025). https://doi.org/10.1007/978-3-031-814 51-8

Ruohomäki, T., Ponto, H., Santala, V., Virtanen, J.P.: Urban digital twin as a socio-technical construct. In: Handbook of Digital Twins, pp. 308–320 (2024). CRC Press. https://doi.org/10.1201/9781003425724

Schrotter, G., Hürzeler, C.: The digital twin of the city of Zurich for urban planning. PFG J. Photogram. Remote Sens. Geoinf. Sci. 88(1), 99–112 (2020). https://doi.org/10.1007/s41064-020-00092-2

SEDIMARK. https://sedimark.eu/, last accessed 28 Mar 2025

Soe, R.M.: FINEST twins: platform for cross-border smart city solutions. In: Proceedings of the 18th Annual International Conference on Digital Government Research, pp. 352–357 (2017). https://doi.org/10.1145/3085228.3085287

Soe, R.M., Ruohomäki, T., Patzig, H.: Urban open platform for borderless smart cities. Appl. Sci. 12(2), 700 (2022). https://doi.org/10.3390/app12020700

TFDS. https://www.ds4sscc.eu/tfds. last accessed 28 Mar 2025

Villanueva-Merino, A., Urra-Uriarte, S., Izkara, J.L., Campos-Cordobes, S., Aranguren, A., Molina-Costa, P.: Leveraging local digital twins for planning age-friendly urban environments. Cities **155**, 105458 (2024). https://doi.org/10.1016/j.cities.2024.105458

Virtanen, J.P., et al.: Contemporary development directions for urban digital twins. Int. Arch. Photogramm. Remote. Sens. Spat. Inf. Sci. **48**, 177–182 (2024). https://doi.org/10.5194/isprs-archives-XLVIII-4-W10-2024-177-2024

Weil, C., Bibri, S.E., Longchamp, R., Golay, F., Alahi, A.: Urban digital twin challenges: a systematic review and perspectives for sustainable smart cities. Sustain. Cities Soc. **99**, 104862 (2023). https://doi.org/10.1016/j.scs.2023.104862

Automated Planning, Execution, and Re-planning of Terrestrial Laser Scanning in the Built Environment

Thivageran Duraimany and Frédéric Bosché[(✉)]

School of Engineering, University of Edinburgh, Edinburgh, UK
{S2266872,f.bosche}@ed.ac.uk

Abstract. Terrestrial laser scanning (TLS) is commonly used for acquiring dense point clouds, used to generate high-quality 3D models, which supports Building Information Modelling (BIM) and digital twinning. However, not only is careful TLS planning necessary to ensure data completeness while minimising scanning time, but real-world conditions often introduce occlusions that prevent original scan plans from achieving the intended coverage. This study presents a method for determining optimal scanner locations through efficient discretisation of target object surfaces into key-points. In addition, it addresses the challenge of dynamically and efficiently adapting scan plans by evaluating whether new scan locations are necessary when occlusions limit visibility. This research uses 3D surface discretisation (into key-points) as the primary method for efficiently evaluating object 3D surface coverage, used during both initial scan planning and dynamic re-planning. A next-best-view algorithm is applied for overall scan plan generation. The method is validated through simulation using a model from the International Society for Photogrammetry and Remote Sensing (ISPRS) dataset, Helios++, and random insertion of clutter. This work contributes to automating TLS workflows, making them more adaptive to real-world uncertainties. Future work will focus on real-world implementation and integration with robotic scanning systems for enhanced automation.

Keywords: Planning for Scanning · Building Information Model · Terrestrial Laser Scanner · Point Cloud · Virtual Laser Scanning

1 Introduction

Terrestrial Laser Scanning (TLS) has emerged as a critical technology for generating high-resolution 3D point clouds, used extensively in Building Information Modelling (BIM), heritage documentation, and digital twinning [1, 16]. Its importance lies in providing accurate and detailed spatial data that are crucial for constructing precise digital (semantically-rich) 3D representations of built environments, essential for maintenance, renovation, and facility management tasks [10, 15].

However, despite its benefits, effective TLS operation is challenging due to the complexity of built environments, which often leads to incomplete data acquisition,

© The Author(s) 2026
A. Jurelionis et al. (Eds.): BDTIC 2025, LNCE 775, pp. 74–85, 2026.
https://doi.org/10.1007/978-3-032-09040-9_7

prolonged scanning times, and higher operational costs [5]. Traditional TLS planning methods often rely on crude planning strategies, which makes them susceptible to inefficiencies and inadequate coverage, especially in complex or obstructed environments [13].

The primary aim of this research is to develop a comprehensive and automated approach for planning, executing, and, as often required (but hardly discussed in the literature), dynamically re-planning terrestrial laser scanning operations within built environments. Specifically, this research seeks to optimise scanner placements and adapt scanning strategies in response to unforeseen practical issues during execution, such as occlusions and inaccessible locations.

Existing TLS planning studies have used techniques such as voxel-based analysis, visibility estimation, and grid-base discretisation to generate candidate scanning positions, alongside optimisation algorithms such as Next-Best-View (NBV) and Genetic Algorithm [1, 4, 9]. These methods have notably contributed to optimising scanner placement, improving scanning efficiency, and improving data completeness and quality [8, 17]. However, despite these advances, many existing methods still lack robust capabilities for adaptive and dynamic re-planning in real-time scanning scenarios, particularly when initial plans confront unforeseen complexities and environmental uncertainties [8]. Recent methods are beginning to address these gaps by introducing continuous optimisation strategies, BIM-based navigation, and adaptive viewpoint generation [4, 8, 17], although efficient and fully automated dynamic re-planning remains a challenge.

Our research introduces a systematic approach that integrates BIM models (in Industry Foundation Classes (IFC) format) to generate potential scanner locations. We discretize target object surfaces into key-points to evaluate visibility and coverage effectively. Using a NBV algorithm, our method then optimises scanner locations to maximize object coverage efficiently, with predefined LOA and LOD requirements. Unlike most existing research, our method explores adaptive surface discretisation, to provide a good balance between robustness and efficiency for elements of various sizes. Another contribution of this research lies in its capability for dynamic re-planning. When practical conditions diverge from the initial assumptions, our approach automatically adjusts the scanning strategy by evaluating coverage deficiencies and determining necessary additional scans.

The paper is structured as follows: Sect. 2 presented related works, Sect. 3 describes the proposed methods, Sect. 4 presents experimental results, and Sect. 4.3 provides conclusions with recommendations for future research.

2 Related Works

Effective spatial data acquisition has become integral in numerous applications within the Architecture, Engineering, and Construction (AEC) industry, particularly in monitoring construction progress, as-built documentation, and quality control. Effective planning of scan operations, commonly referred to as *Planning for Scanning (P4S)*, has become an essential research focus due to the need to achieve complete, accurate and efficient data collection under various practical constraints [1].

Early approaches to TLS P4S predominantly used voxel-based analysis, visibility estimation, and Next-Best-View (NBV) strategies [1, 8, 17]. These methods provided significant progress in optimising scanner locations, ensuring data completeness, and reducing operational redundancy. Recent advancements have brought about continuous optimisation methods and model-based approaches. Zeng et al. [17] proposed a novel continuous multi-objective optimisation framework to determine optimal scanning locations, reducing the computational complexity typically encountered in discretization-based methods. This approach leveraged a breadth-first strategy combined with a back-tracking algorithm to systematically reduce the number of scan locations, improving both the completeness of the data and the planning efficiency. However, this method does not address the computational inefficiencies resulting from high-resolution discretisation, nor does it incorporate real-time adaptation after each scan to evaluate data sufficiency.

The automation of adaptive scan planning in complex 3D environments was further explored by Noichl et al. [8]. Their method introduced automated mesh processing, viewpoint candidate generation, and advanced visibility and coverage evaluations to handle dynamically changing requirements and conditions. This work demonstrated significant improvements over manual expert-generated scan plans, highlighting its practical applicability for complex indoor environments. Nonetheless, it does not consider scan location accessibility, particularly in cases where certain locations may be inaccessible or obstructed in real-world scenarios.

Integration with BIM models represents another substantial progression in TLS scan planning. Frías et al. [4] developed an automated scan planning system based on BIM models, employing visibility analyses and optimization techniques such as Ant Colony Optimization (ACO) for route planning. Their approach showed improved scanning efficiency, which is particularly beneficial for automated systems mounted on mobile robotic platforms. Similarly, Park et al. [9] expanded on BIM-based methods by introducing a systematic framework specifically tailored for quadruped robotic platforms, optimizing both scan positions and scan ordering within navigable spaces derived from BIM models. This approach also demonstrated notable reductions in scanning time and scan positions compared to an expert-generated plan.

Despite these advancements, the comprehensive review by Aryan et al. [1] highlighted ongoing challenges in achieving fully automated and adaptive scan planning. Specifically, it emphasised the need for solutions that can dynamically re-plan scan operations in real-time as environmental conditions evolve.

3   Method

The proposed method for generating an optimised scan plan and performing dynamic scan re-planning is divided into two major phases. The first phase focuses on Scan Plan Creation, which is depicted in the top flowchart and detailed in Sect. 3.1. The second phase addresses Dynamic Scan Re-Planning during scanning execution as illustrated in the bottom flowchart and detailed in Sect. 3.2. This phase is crucial for adapting the scan plan in response to practical challenges encountered during the scanning operation (Fig. 1).

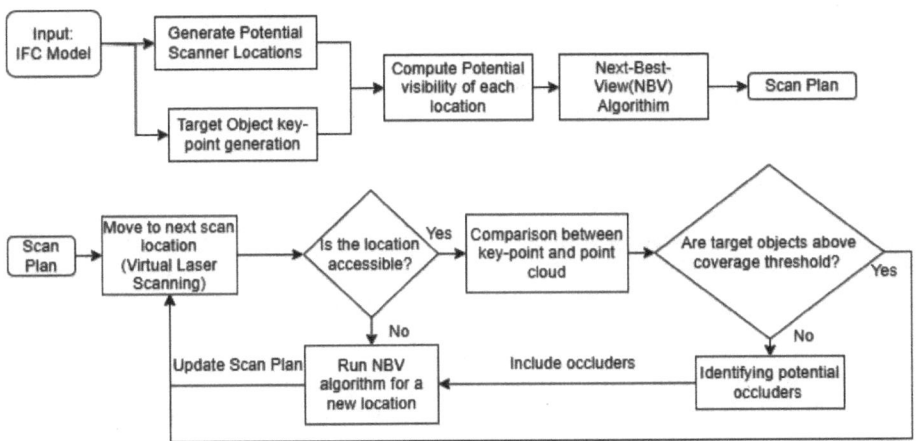

Fig. 1. Overview of the proposed method.

3.1 Scan Plan Creation

This paper introduces a systematic approach to generating an optimized scan plan for TLS using a prior Building Information Model (hereafter 'BIM') model. In particular, the approach uses an IFC representation of the BIM model as input, providing architectural and structural spatial data. Potential scanner locations are derived on the basis of the geometry of the floor. Concurrently, the surface of each target object is discretized into 'key-points'. Then, the visibility of these key-points from each potential scanner location is quantitatively evaluated. Using a NBV algorithm, an optimal set of scanner positions is determined, ensuring maximum coverage of the Objects of Interests with minimal redundancy. The optimized scan plan facilitates efficient and comprehensive data acquisition. The methods used for these steps are elaborated in the following subsections.

Generate Potential Scanner Locations. In this step, the floor geometry is extracted from the IFC model to establish a baseline for scanner placement.

A structured square grid is generated across this extracted floor surface, with individual grid points uniformly spaced at a resolution of $r = 1$ m. Each point in the grid represents a hypothetical position of the terrestrial laser scanner. For consistency and comparability within this study, the height of each scanner is fixed at 1.2 m above the floor plane. Subsequently, a filtering step is applied to these hypothetical positions by employing an expanded bounding box around the BIM elements. Any scanner positions falling within this expanded bounding box are discarded. This filtering process ensures that the set I of remaining positions represents practical and viable scanner locations within the given environment.

Target Object Key-Point Generation. Given that the objects are represented by 3D meshes composed of multiple triangles, all computations are performed at the triangle level. This subsection focuses on generating key-points to effectively represent objects with sufficient coverage while optimizing computational efficiency. Previous work has considered various forms of surface discretisation to establish the level of coverage of elements. Dense discretisation gives more robust results, but at increased computational costs. To balance computational load and accuracy, Poisson disk sampling with a spacing parameter of $r = 0.25$ m is used to uniformly sample points on the surface of the triangles that make up the 3D mesh of each target object. We refer to those points as *key-points* representing the surface of the object, where $K_{o,i}$ denotes set of key-points for each object and each triangle. The sampling distance yields a fairly space key-point discretisation, which we find works well for sufficiently large triangles, but not smaller ones. So, if a triangle has three or fewer points after Poisson sampling, then the set is discarded and replaced with the set of four key-points, comprising its three vertices and its centre point. This ensures consistent surface representation across all triangle sizes.

Compute Potential Visibility of Each Location. The objective of this step is to evaluate how effectively each scanner location can visually capture the target objects, and more specifically their key-points. A key-point is considered visible from a given scanning location if:

Distance: the point is less than a maximum distance from the scanner ρ_{max} set given the specifications of the selected scanner and expected scanning Level of Accuracy (LOA). Incidence Angle: the incidence angle between the ray and the normal of the object's surface at the key-point location is less than a maximum threshold α_{max}.
Occlusion: there is no occlusion by BIM model object between the scanner and the key-point.

This occlusion check requires the use of a ray-triangle intersection method to check the intersection of the bounded ray (from the scanner location to the key-point) with all triangles of all BIM objects. For this, we use the Möller-Trumbore intersection algorithm [7]. Because this intersection test is computationally intensive | due to the total numbers of elements and corresponding triangles in the BIM model [14] | the Distance and Incidence Angle checks are conducted first. For example, in this study we use the characteristics of a BLK-360 laser scanner, limiting the incidence angle to a maximum of $75°$ and the maximum range to 20 m.

The outcome of this step yields a detailed set of the visibility of each key-point, $\bar{v}_k$ of each triangle of each target object, for each potential scanner location. By aggregating this information across all scanning locations, it is possible to infer the *theoretically visible area* of each triangle $\bar{a}_{o,i}$ and then each object $\bar{a}_o$.

$$\bar{a}_{(o,i)} = \left(\frac{\sum_{k \in \mathcal{K}_{(o,i)}} \bar{v}_k}{|\mathcal{K}_{(o,i)}|} \right) \times a_{(o,i)} \tag{1}$$

$$\bar{a}_{(o)} = \sum_{i \in \mathcal{I}} \bar{a}_{(o,i)} \tag{2}$$

Next-Best-View (NBV) Algorithm. The goal of this step is to now select the smallest set of scanner locations that achieves a minimum (set by the user) coverage of the surfaces of target objects. Similar to other researchers, we adopt the NBV algorithm [12]. The NBV algorithm prioritizes locations that provide maximum additional visibility to previously unobserved or minimally observed objects whose total observed surface remains below the minimum threshold set by the user. An essential aspect of this step involves setting the minimum visibility threshold. We set this threshold as a percentage of the theoretically visible surface of each target object. In this research, an object is considered to have sufficient coverage when it attains visibility of at least $\tau = 65\%$ of its theoretically visible surface area.

3.2 Dynamic Scan Re-planning

The set of scanning locations obtained above can now be used by the scanning agent (human or robot), who must now visit each location to acquire the expected data. However, the situation on site may differ from that assumed during the planning phase. The dynamic scan re-planning process aims to adaptively respond to such unexpected conditions and challenges encountered during the scanning operation. These can include the impossibility to access a given scanning location, or unexpected occlusions preventing the scanning of parts of an object as originally planned.

The core concept of dynamic re-planning is to perform systematic checks after each performed scan to (1) verify whether the acquired data achieve the intended coverage as per the plan, and (2) if the data are deemed insufficient, the system automatically adjusts the scan plan. These two steps are detailed in the following corresponding sub-sections.

Comparison of Expected and Actual Coverage. To verify whether the point cloud data acquired from the current scanning location achieves the intended coverage as per the plan, we evaluate whether the acquired point cloud "covers" the key-points expected to be acquired from that location.

For each expected key-point, the closest points in the point cloud acquired on site are found. This is achieved by performing a nearest-neighbour search using a hybrid algorithm facilitated by a k-dimensional tree (k-d tree) data structure [2]. The process utilizes two primary parameters: a distance threshold $d_{\text{threshold}}$ and a fixed number of maximum neighbours to return $n_{\text{Neighbors}}$, which together determine the visibility criterion.

1. Perform a nearest-neighbour search around each key-point $\mathbf{k}_i$, constrained within radius $d_{\text{threshold}}$, retrieving n closest points from P.
2. A key-point is considered *covered* if $n \geq n_{\text{Neighbors}}$.

After completing the comparison between key-points and the generated point cloud, the total visible area of each target object is re-evaluated. This re-assessment determines whether any target object's actually covered surface (from all scans actually acquired up to the current location and all remaining planned scans) falls below the designated coverage threshold of τ (65%). If no object is found to have its actual covered surface fall below τ, then the agent is instructed to simply proceed to the next scan location. However, if an object is found to have its actually covered surface fall below τ, then the system explores whether one or more additional locations can be found to acquire additional data that would allow the acquired object surface to once again get above the threshold τ, the process for which is described in the following section.

Selecting Additional Scanning Location. The process works but can be very slow (because it is 'dumb') when occlusions are encountered. Indeed, we have observed that, when new scanning locations are required due to occlusions, the system tends to suggest locations right next to the original one, even though the same issues are likely faced for them. To address this, we propose to dynamically model the occluders and evaluate their impact on the new location suggested by the system before actually moving to that location.

Identifying Potential Occluders. We find and model the occluders with meshes, as follows:

1. We compute (offline) an extended Oriented Bounding Box (xOBB) around each BIM object.
2. The points from the last scan that fall outside this bounding box are considered potential occluding objects.
3. The points outside the objects' bounding boxes are voxel-downsampled to reduce computational load, and statistical outlier removal techniques [11] are employed to eliminate noisy data points typical in laser scanning datasets.
4. Density-Based Spatial Clustering of Applications with Noise (DBSCAN) algorithm [3] is applied to cluster the remaining points systematically. Each result point cluster is considered to be an occluder. DBSCAN clustering requires two critical parameters: epsilon (ϵ) and the minimum number of samples. Given the consistent point density produced by the simulated laser scanner, the minimum number of samples is fixed. To efficiently determine an optimal value of epsilon, a k-distance graph is utilized.
5. For each occluder, Oriented Bounding Boxes (OBBs) are generated around its points, clearly delineating them as occluding 3D meshes.

Location Selection given Occluders. This subsection details the process of selecting optimal scanner locations when considering modelled occluders within the environment. The primary objective is to identify new scanner positions that maximize the visibility and coverage of target objects whose visibility falls below τ. To achieve this, Next-Best-View (NBV) analysis is performed to systematically evaluate all potential locations based on the visibility area coverage of each object [12]. To accurately incorporate occlusion effects, each candidate scanner location undergoes a rigorous validation step. Specifically, visibility of each key-point on the target object is individually assessed by employing the Möller–Trumbore ray-triangle intersection algorithm [7]. This algorithm efficiently determines whether a direct line of sight from the scanner to each key-point intersects with any triangle from the occluder's mesh model. If a key-point is determined to be occluded (i.e., intersected by an occluder triangle), it is excluded from the visibility computation at that location.

Following this assessment, the A_{vis} is recalculated for each candidate location. Finally, the location that offers the highest visible area coverage, after accounting for occlusions, is selected as the optimal scanner position for the given target object. This methodological integration of the Möller–Trumbore intersection algorithm ensures that the resulting scanner placements accurately reflect practical constraints and real-world occlusion scenarios [7].

4   Experiments

This section presents the results obtained when applying the proposed method on a benchmark dataset. The effectiveness of the scan plan is first evaluated through visual inspection, showing the distribution of scanner locations optimized via the NBV algorithm. The method is applied to the International Society for Photogrammetry and Remote Sensing Benchmark dataset, particularly the Technische Universität Braunschweig, Germany (TUB1) IFC model [6].

The scanner used for this experiment is the BLK 360 laser scanner. It has a maximum scanning distance of 20 m, LOA $\approx \pm 4$ mm and LOD ≈ 51 mm.

4.1   Initial Scan Plan

The scan plan generated by the proposed method is presented in Fig. 2. The scanner locations are represented by red markers that highlight their strategic placement to ensure maximum coverage. The 182 scanner locations, denoted by $\mathscr{I}$, are evenly distributed 1 m apart, avoiding unnecessary redundancy and ensuring that all target objects achieve theoretical visibility τ (65%).

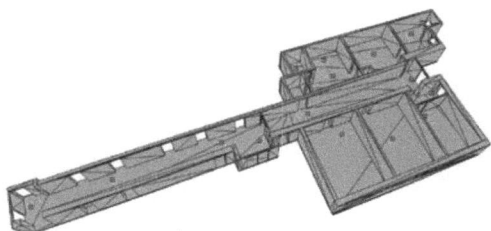

Fig. 2. Optimised set of scanner locations within the IFC model environment.

4.2   Scan Plan Execution and Replanning

Due to the fact that the authors do not have access to the selected building, the evaluation of the Scan Plan Execution and Replanning process is conducted through simulation. For this, we assume that the scanning agent (human or robot), position themselves to each planned scanning location and we perform a virtual laser scanning operation using the Helios++ package. This simulation mimics the process of a surveyor or a robotic system performing the scanning task at each predefined scanner location.

To evaluate the dynamic re-planning framework, we show the example when an occluder artificially introduced into the environment, as shown in Fig. 3a.

Occluder Detection and Processing.
In the situations depicted in Fig. 3a, an issue arises when the actual visibility of the target wall behind the introduced occluder falls below the predefined coverage threshold of τ = 65%. This visibility drop indicates a blockage or obstruction at the planned scanner location, prompting further analysis.

The acquired point cloud is thus further analysed to determine the shape and location of the occluder. Figure 3b illustrates the point cloud obtained from the virtual scan performed at the indicated location, clearly showing the occluder situated between the scan position and the wall. Figure 3c shows the isolated occluder point cloud, which is extracted by removing all points corresponding to known IFC model elements using an expanded bounding box approach (as described in Sect. 3.2). The red points are outlier points that are flagged for removal based on the DBSCAN clustering process. Figure 3d shows the initial IFC model with the modelled detected occluder.

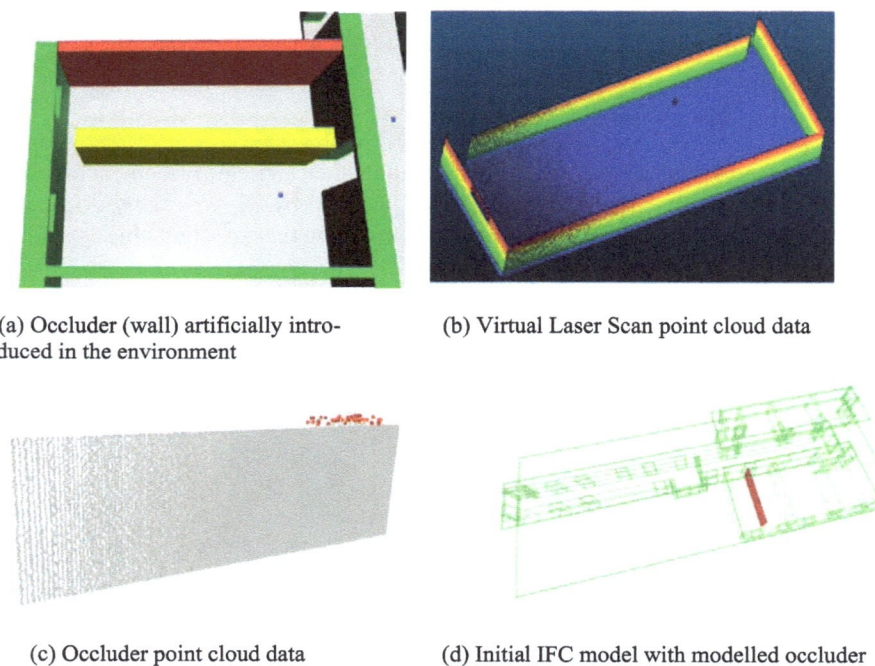

(a) Occluder (wall) artificially introduced in the environment

(b) Virtual Laser Scan point cloud data

(c) Occluder point cloud data

(d) Initial IFC model with modelled occluder

Fig. 3. Visualization of occluder identification and processing

Generation of New Scanner Location.

After identifying and modelling the occluder, the NBV algorithm is executed again to select one (or more) additional scanner location(s), aimed at mitigating the visibility loss caused by the occlusion. Figure 4 demonstrates the result of this dynamic re-planning step. In Fig. 4, the blue point indicates the initial planned scanner location from where the occluder was detected, and the red point shows the newly generated additional scanner position.

The result of the selection of this new additional scanning location is that it restores the overall coverage of the wall behind the occluder to a value of 70%, which is above the threshold of $\tau = 65\%$. The scanning agent can then proceed to the next originally planned scanning location.

Fig. 4. Visualization of the dynamically generated new scanner location (red point), replacing the original scanner location (blue point) to resolve visibility issues caused by the occluder (yellow).

4.3 Conclusion

This paper presented the effectiveness and adaptability of the proposed scan-planning and dynamic re-planning methodology. Initially, the scan plan provided by the NBV algorithm successfully identified the optimal scanning positions based on the IFC model, effectively maximizing theoretical visibility coverage, while accounting for LOA and LOD specifications. However, practical considerations such as occlusions and location accessibility, simulated within the virtual scanning environment, require real-time adjustments to maintain data completeness and scan quality.

The dynamic re-planning approach demonstrated its robustness in identifying and resolving visibility losses due to unforeseen occlusions. By efficiently detecting and modelling occluding objects from the virtual point cloud data, the method could dynamically recompute optimal scanning locations. The results indicated that this re-planning successfully recovered lost coverage, ensuring target objects consistently met the visibility threshold.

Future experiments will explore the impact of approximate scanner location during execution, and more extensive comparison with existing P4S methods.

Acknowledgements. We would like to thank Petroliam Nasional Berhad (PETRONAS) for their financial support for this research. For the purpose of open access, the authors have applied a Creative Commons Attribution (CC BY) licence to any Author Accepted Manuscript version arising from this submission.

References

1. Aryan, A., Bosché, F., Tang, P.: Planning for terrestrial laser scanning in construction: a review. Autom. Constr. **125**, 103551 (2021). https://doi.org/10.1016/j.autcon.2021.103551

2. Bentley, J.L.: Multidimensional binary search trees used for associative searching. Commun. ACM **18**(9), 509–517 (1975)
3. Ester, M., Kriegel, H.P., Sander, J., Xu, X., et al.: A density-based algorithm for discovering clusters in large spatial databases with noise. KDD **96**, 226–231 (1996)
4. Frías, E., Díaz-Vilariño, L., Balado, J., Lorenzo, H.: From BIM to scan planning and optimization for construction control. Remote Sens. **11**(17), 1963 (2019)
5. Kersten, T.P., Lindstaedt, M.: Image-based low-cost systems for automatic 3d recording and modelling of archaeological finds and objects. In: Progress in Cultural Heritage Preservation: 4th International Conference, EuroMed 2012, Limassol, Cyprus, October 29–November 3, 2012. Proceedings 4, pp. 1–10. Springer (2012)
6. Khoshelham, K., Díaz Vilariño, L., Peter, M., Kang, Z., Acharya, D.: The ISPRS benchmark on indoor modelling. Int. Arch. Photogram. Remote Sens. Spatial Inf. Sci. XLII-2/W7, 367–372 (2017). https://doi.org/10.5194/isprs-archives-XLII-2-W7-367-2017. https://isprsa rchives.copernicus.org/articles/XLII-2-W7/367/2017/
7. Möller, T.: A fast triangle-triangle intersection test. J. Graph. Tools **2**(2), 25–30 (1997)
8. Noichl, F., Lichti, D.D., Borrmann, A.: Automating adaptive scan planning for static laser scanning in complex 3D environments. Autom. Constr. **165**, 105511 (2024)
9. Park, S., Yoon, S., Ju, S., Heo, J.: Bim-based scan planning for scanning with a quadruped walking robot. Autom. Constr. **152**, 104911 (2023)
10. Pătrăucean, V., Armeni, I., Nahangi, M., Yeung, J., Brilakis, I., Haas, C.: State of research in automatic as-built modelling. Adv. Eng. Inform. **29**(2), 162–171 (2015)
11. Rusu, R.B.: Semantic 3d object maps for everyday manipulation in human living environments. KI-Künstliche Intelligenz **24**, 345–348 (2010)
12. Scott, W.R., Roth, G., Rivest, J.F.: View planning for automated threedimensional object reconstruction and inspection. ACM Comput. Surv. (CSUR) **35**(1), 64–96 (2003)
13. Thomson, C., Apostolopoulos, G., Backes, D., Boehm, J.: Mobile laser scanning for indoor modelling. ISPRS Ann. Photogram. Remote Sens. Spatial Inf. Sci. II-5/W2, 289–293 (2013). https://doi.org/10.5194/isprsannals-II-5-W2-289-2013. https://isprsannals.copernicus.org/art icles/II-5-W2/289/2013/
14. Wald, I., Havran, V.: On building fast KD-trees for ray tracing, and on doing that in o (n log n). In: 2006 IEEE Symposium on Interactive Ray Tracing, pp. 61–69. IEEE (2006)
15. Willkens, D.S., Liu, J., Alathamneh, S.: A case study of integrating terrestrial laser scanning (TLS) and building information modeling (BIM) in heritage bridge documentation: the edmund pettus bridge. Buildings **14**(7), 1940 (2024)
16. Wu, C., Yuan, Y., Tang, Y., Tian, B.: Application of terrestrial laser scanning (TLS) in the architecture, engineering and construction (AEC) industry. Sensors **22**(1), 265 (2021)
17. Zeng, Y., et al.: Optimal planning of indoor laser scans based on continuous optimization. Autom. Constr. **143**, 104552 (2022)

Digital Twins for Data Centre Cooling Optimisation and Waste Heat Recovery

Sara Giordani, Rossano Scoccia$^{(\boxtimes)}$, and Marcello Aprile

Department of Energy, Politecnico di Milano, Milan, Italy
rossano.scoccia@polimi.it

Abstract. The paper explores the development of a Digital Twin (DT) for the management and optimisation of heat recovery in data centres, using HYCOOL-IT project demo site. The project features the development of a "Building Digital Twin Environment" as a "Platform as a Service", integrating SIMBOT-based interactive simulators and Web API microservices to support processes from planning to performance evaluation. The article describes the integration of a waste heat recovery system within the data centre Z3 to heat the adjacent university building BL26. The DT architecture incorporates real-time data processing and a Software-in-the-Loop Model Predictive Control system to optimise control actions. Two DTs are under development: the data centre DT and the active heat recovery system DT. The former will optimise performance by continuously monitoring and controlling data centre conditions, cooling system settings, and power supply system. Simulations of the server room, chillers, and power systems will enable testing of management strategies for use cases like free cooling and emission subsystem operation optimisation. The latter will monitor the integration of a water-water heat pump (WW-HP) with the BL26 heating system, tracking electricity use, water temperatures, and flow rates. Control strategies will adjust setpoints, operation modes, and flow or temperature differences to improve system efficiency. Dynamic simulations could assess BL26 heating demand and WW-HP. Use cases include integrating WW-HP into the current plant, optimising mid-season operation, and implementing free cooling strategies to enhance energy efficiency.

Keywords: Digital Twin · Data Centre · Waste Heat Recovery · Energy Efficiency · Heat pump · BDTE · PaaS · MPC · SiL

1 Introduction

The concept of the Digital Twin (DT) was first introduced in 2002 by Michael Grieves at the University of Michigan in the context of Product Lifecycle Management. The proposed model comprised three key components: a real space, a virtual space, the link for data flow from real space to virtual space and the link for information flow from virtual space to real space [1].

Although several authors and researchers have attempted to define the concept of DT for the construction sector, currently there is no universally accepted definition of the term [2]. Recently, a paper [3] has derived the definition of Digital Twin in the built

© The Author(s) 2026
A. Jurelionis et al. (Eds.): BDTIC 2025, LNCE 775, pp. 86–98, 2026.
https://doi.org/10.1007/978-3-032-09040-9_8

environment from over 15,000 scientific publications, highlighting significant variations in definitions based on applications domains. A DT is often described as a continuously updated virtual replica of a physical entity [4]. The connection between the virtual-physical duality is provided by data in its various forms [5]. The way this data circulate between digital models and physical components is one of the key points of DT definition. Some consider a digital model to be a DT if it incorporates real-time data, enabling continuous monitoring of the physical entity actual operation. Others, however, believe that a system can only be considered a DT if it captures and streams data to a digital platform that performs real-time analysis to optimise the design and the performance [6]. In this view, real-time data must be processed to generate feedback that enhances the management and operation of the physical entity. Hence, application of DT includes real-time monitoring, designing/planning, optimisation, maintenance, remote access, etc. [7]. Data flow between the digital and physical twins are continuously updated and adapted to the changes in the physical asset – supported by technologies such as AI, machine learning, sensors and IoT. This allows sharing insights, supporting decision-making, and enabling simulation, prediction, monitoring, control, and performance optimisation of the physical asset throughout its lifecycle [8]. In the case of bidirectional information exchange, a crucial point of discussion is how data is transmitted from the digital model to the physical entity. Data transmission can either be fully automated within the virtual-physical loop or involve a human intermediary who evaluates whether to implement the changes suggested by the optimisation process. A committee is currently drafting a standard [9] for this approach, and we are aligning our work with their proposed definitions: "Digital representation of a target entity with data connections that enable convergence between the physical and digital states at an appropriate rate of synchronization. Digital twin has some or all the capabilities of visualization, simulation, surveying, monitoring, forecasting, control, etc.".

DTs have the potential to significantly enhance the Operation and Maintenance (O&M) phase of building assets [10]. They enable more efficient use of O&M data, improve the perception and visualisation of building systems, and facilitate automated feedback control [11]. DT-based anomaly detection process allows for continuous monitoring of building components, thereby enhancing automated asset monitoring during the O&M phase [12]. DT technologies also support more efficient and responsive Facility Management (FM) planning and control by providing real-time status updates of building assets, ultimately improving the performance of MEP systems throughout the O&M phase [13]. However, challenges remain in real-time decision-making, data interoperability, and predictive analytics, highlighting the need to bridge the gap between theoretical advancements and practical, real-world applications [14].

Despite these promising capabilities, several challenges still hinder the widespread adoption of DT solutions. Many of these challenges stem from the novelty of the technology, resulting in a lack of standards and regulations, a shortage of competent engineers and technicians, and limited availability of supporting software [7]. Additionally, the effective deployment of DT solutions depends heavily on robust data acquisition systems and their integration with Building Management Systems (BMS), which often requires specialized engineering expertise [15]. Moreover, the quality of collected data – serving as input for analyses and simulations on the DT virtual entity – remains critical to ensure the accuracy and reliability of the results [16]. Nevertheless, despite these barriers, the integration of DT technologies continues to expand, particularly in the domains of thermal comfort and energy management for buildings, with an increasing number of studies focusing on thermal comfort monitoring, visualization, tracking, energy management, prediction, and optimization for existing buildings [17].

This paper concerns the DT applications for a data centre and its waste heat recovery system. Waste Heat Recovery (WHR) technology is considered as a promising approach to improve energy efficiency, achieve energy and energy cost savings, and mitigate environmental impacts [18]. The DT environment is implemented using the IDP [19] platform, which facilitates the connection of physical devices to the cloud, allowing data collection and real time access.

2   Case Study Description

2.1   HYCOOL-IT Project

HYCOOL-IT is a three-year research and innovation initiative currently in its 16th month, bringing together nine partners from six European countries. The main objective is to develop processes supported by innovative solutions, both digital and technical, for the efficient and reliable implementation of IT server rooms in advanced tertiary buildings, with a particular focus on replicability through standardization. HYCOOL-IT includes the creation of a Building Digital Twin Environment (BDTE), developed as a Platform as a Service (PaaS), with specific Web API microservices to connect dedicated tools supporting planning, design assessment, commissioning, and performance evaluation. This includes the so called SIMBOTs to facilitate the effective integration and operation of these advanced server rooms within buildings. A SIMBOT is an open-source, semantically structured representation – or a corresponding library component within a simulation environment – of the mathematical model of a commercial equipment. Built using standardised library components, mathematical functions, fluid dynamics, and interface ports, the model is designed for real-time execution and aligns with the manufacturer's performance specifications. The implementation could be done using FMI standard [20]. Moreover, SIMBOTs pave the way for the future prescription of the innovative technical equipment proposed in the project. In addition, existing rooms can be mirrored and enhanced with the usage of BDTE to generate improved baselines and forecasting, supporting designers in the design phase by helping them choose the best design option. Maintenance engineers can also leverage BDTE to enable performance contracting and ensure the realization of key performance indicators (KPIs).

All methodology and software solutions will be tested and validated in the Bovisa Campus at the Z3 data centre, undergoing renovation, serving as a representative TRL 5 Living Lab. This paper will show the analysis conducted on the Z3 data centre and the WHR system designed to recover heat from the server room to heat up the adjacent building. It will also discuss the potential benefits of developing a DT for both the data centre and the heat recovery system.

2.2 Data Centre Z3

"Z3" is the nickname of the data centre of the University Politecnico di Milano completed in 2013. It is a small data centre with a net climatized area of around 150 m². The cooling load in the starting phase was around 80 kW. The forecasted cooling load at maximum expansion is 320 kW. The current cooling load is 190 kW. The servers are installed in four rows of racks and grouped in two HAC (Hot-Aisle Containment) islands. One HAC island is for university administrative services, while the other is for High Performance Computing (HPC). Figure 1 displays the schematic floor plan of Z3 data centre, complete with description of the various components and zones.

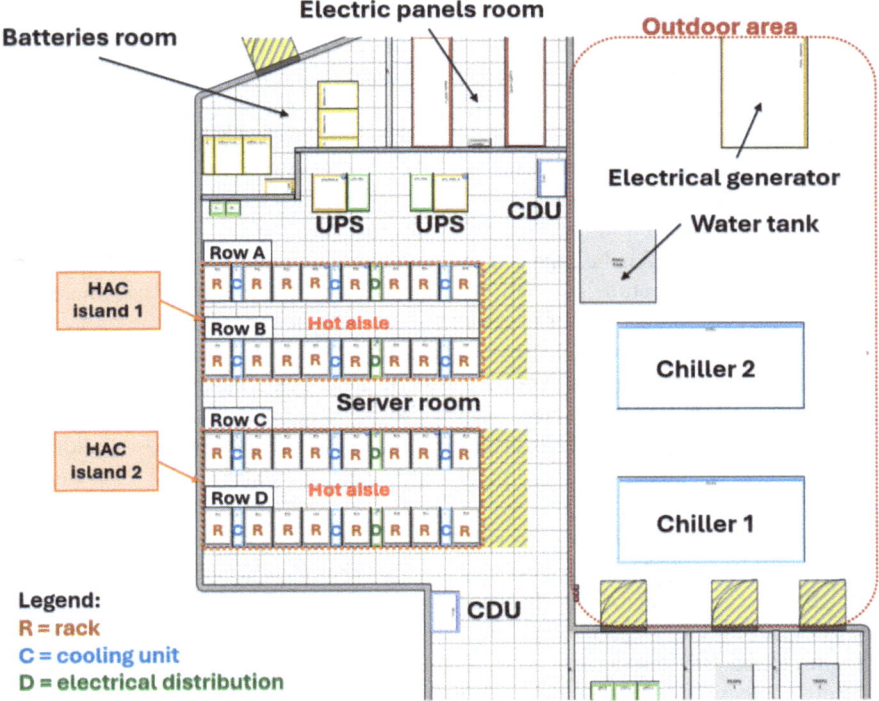

Fig. 1. Z3 data centre schematic floor plan.

The Z3 cooling plant is composed of two chillers, two cooling distribution units (CDUs) and twelve InRow Half Rack RC terminals ("InRow cooling units" from now on), they are a kind of fan-coil cooling emission sub-system. The indoor part of the data centre is divided in three spaces: the server room, the battery room, and the electric panels room. The server room hosts the server racks, the InRow cooling units, the CDUs and the uninterruptible power supply units (UPSs). The outdoor part of the data centre hosts the two chillers, the water storage tank, and the electrical generator.

To ensure cooling plant redundancy, the chillers operate alternately. An energy audit is conducted on the data centre Z3, revealing that the cooling system is not working efficiently. In fact, the Energy Efficiency Ratio (EER) is lower than expected. During warmer months is just around 1.75, while it increases to approximately 2.9 during colder months. The heat produced by the servers is captured by the InRow cooling units, which re-emit cooled air into the room.

Currently, the heat produced by the servers and balanced by the InRow cooling units is dissipated without any recovery. Therefore, a system that can recover the heat produced by the servers to heat an adjacent building is designed.

2.3 Waste Heat Recovery System

Two buildings of the Bovisa campus, located near the data centre, are analysed to determine the most suitable location for the heat recovery system: Building BL26 and Building BL27. For each building, a detailed study of the HVAC system is conducted, along with a comprehensive energy audit and performance assessment. For the building energy performance assessment, the RELAB app [21, 22] is used. The RELAB app is a software based on the EN ISO 52016 standard [23], where the building characteristics can be defined to estimate the energy demand. Based on the analysis, Building BL26 is chosen to host the heat recovery system.

The most straightforward approach for reusing waste heat from the server room is to transfer it to the building heating system via a water-to-water heat pump (WW-HP) to increase the heat temperature level. The heat recovery system will involve the installation of a heat exchanger in the outdoor plant area of Z3 and a heat pump on the roof of BL26. The heat exchanger will hydraulically separate the server room cooling system from the waste heat recovery system. The primary side of the heat exchanger will be connected to the cooling system in the server room while the secondary side will be connected to the heat pump of building BL26 heating system. A schematic representation of the waste heat recovery system and its connection to Z3 data centre and BL26 is displayed in Fig. 2.

Careful consideration must be given to selecting a water-to-water heat pump and integrating it into both the server room cooling system and the existing heating system. Indeed, the heat pump capacity is chosen in function of both the cooling load of the server room and the building peak heating demand. Since a heat pump heating capacity varies with source and sink temperatures, selection is based on performance at the required operating temperatures, rather than nominal capacity. The heat pump shall have an inlet temperature from the evaporator side of 15 °C and an outlet temperature from the evaporator side of 10 °C. The flow temperature in the system circuit of the BL26 shall be 58 °C while the return temperature shall be 50 °C. It is important that the heat pump model chosen for installation has good capacity control characteristics, so selecting multi-compressor models with 3–4 capacity stages is important to improve performance under partial load and offer scalability.

A preliminary analysis indicates that integrating a water-water heat pump into BL26 heating system could reduce natural gas consumption and lower overall energy costs by approximately 15%, while also decreasing CO_2 emissions by around 23%. The implications related to the waste heat recovery plant will be thoroughly discussed in a future work.

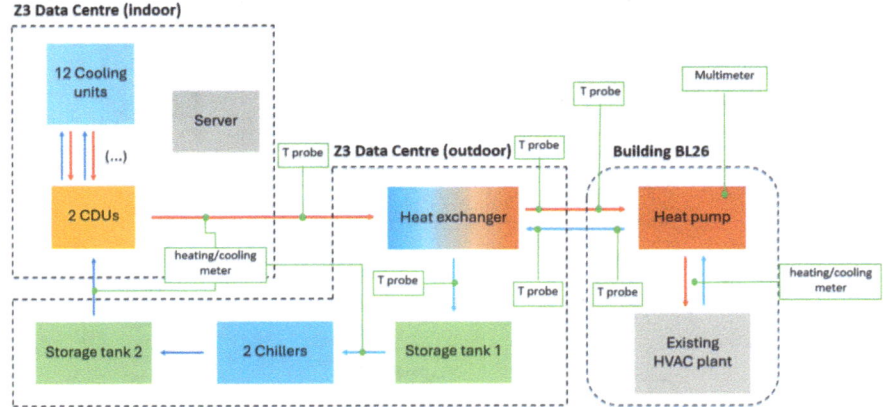

Fig. 2. Waste heat recovery plant scheme.

3 Building Digital Twin Environment Ongoing Development

The following chapters detail the architecture of the DT, with a focus on real-time data processing and the key features of the DT environment.

3.1 Real-Time Data Processing

In the Digital Twin architecture, the integration between physical devices and cloud-based analytics is essential for real-time monitoring and optimisation. The IoT HUB Service plays a central role in this process, enabling efficient data collection, transmission, and processing. It is an integrated service within IDP's DT platform, allowing the connection of physical devices with the cloud so that the data obtained by the devices or sensors can be visualised and consulted in real time from any type of client application that connects with the IoT HUB service through a web API. Physical devices equipped with sensors continuously capture operational parameters, which are then transmitted using protocols such as MQTT, AMQP, and HTTPS. The data is sent to an IoT HUB (Fig. 3), where it is processed and then forwarded to both a web API and an SQL database for analysis and storage. By structuring and storing sensor data in a centralised system, the Digital Twin can leverage it for simulation and predictive analysis, optimising system performance and enabling proactive decision-making.

One of the most relevant goals of a Digital Twin is to allow continuous and automatic system optimisation. This architecture ensures a continuous feedback loop between the physical and digital layers, enhancing operational efficiency, improving maintenance strategies, and supporting energy optimisation efforts.

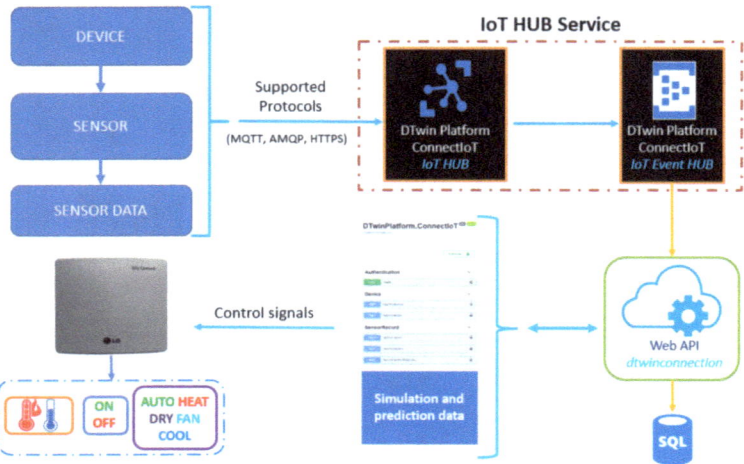

Fig. 3. IoT HUB integration for real-time data processing in the DT architecture: schematic representation.

3.2 Digital Twin Environment Development

The development of a DT environment (Fig. 4) for data centre optimisation integrates real-time monitoring, simulation models, and predictive control mechanisms to enhance operational efficiency. The system collects real-time data on server room conditions, InRow cooling settings, chiller operations, and power supply performance. These data inputs are processed by the Simulation Model Tracking System (SMTS), along with the synthetic data generated by mathematical models (SIMBOTS), which simulate the behaviour of chillers, power supply systems, and the server room to provide a virtual representation of the physical infrastructure. A key component of this architecture is the Software-In-the-Loop Model Predictive Control (SiL-MPC) system, which employs predictive modelling and optimisation techniques to determine the most efficient control actions. By continuously optimising control trajectories throughout the year and incorporating forecasts of boundary conditions – such as weather data – the system enables more informed and energy efficient operational decisions. The SiL-MPC module processes current system data, simulates future states, and applies optimised control strategies to chillers, InRow cooling systems, and the overall server room environment. This continuous feedback loop ensures energy efficiency, stability, and proactive system optimisation, ultimately enhancing the reliability and performance of critical infrastructure in data centres.

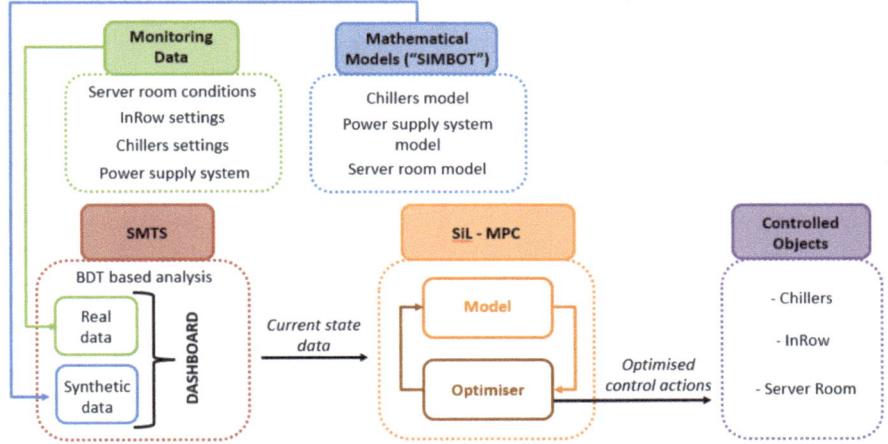

Fig. 4. Data flow in the DT environment: Monitoring Data, Mathematical Models, SMTS and SiL-MPC schematisation.

An essential component of the DT environment is the dashboard, which serves as a centralised interface for visualizing and analysing data collected by the Simulation Model Tracking System (SMTS). The dashboard integrates multiple functionalities to enhance data analysis and decision-making. It includes a viewer, which offers an interactive 3D representation of the physical infrastructure, allowing users to navigate through the simulated environment. The monitoring section displays operational parameters, facilitating real-time assessment of performance. Additionally, the KPI section presents processed data in a graphical format, allowing to interpret the complex system behaviours. In particular, it can display performance metrics derived from simple monitoring data, such as Power Usage Effectiveness (PUE) and Energy Efficiency Ratio (EER), providing valuable insights into energy efficiency and operational effectiveness. To further enhance collaboration, the shared documents section enables centralized document management, ensuring seamless access to relevant information.

4   Applications

4.1   Data Centre Digital Twin

The DT of the data centre environment relies on a set of measured, controlled, and simulated variables to optimise performance and energy efficiency. Sensors located in Z3 data centre measure several parameters that can be checked on EcoStruxure platform (by Schneider Electric) [24] – an infrastructure for cloud-connected digital services. These parameters concern the server room conditions, the InRow settings, the chillers settings, and the power supply system.

The data centre cooling system control variables are:

1. For the InRow cooling units:
   a. room air setpoint temperature – from 18.0 °C to 32.2 °C;

b. supply air setpoint temperature – from 15.0 °C to 30.2 °C;
2. For the chillers: supply water setpoint – from 7 °C to 13 °C.

Through simulations, the behaviour of chillers, the server room, and the power supply system will be reproduced. This will allow different management strategies to be tested without affecting regular data centre operations, enabling the identification of the most efficient solutions.

Data analysis and simulations are applied to various use cases, such as free cooling optimisation and InRow operation optimisation, both in terms of energy efficiency and acoustics. Indeed, in this specific project also acoustic comfort is considered because there are often system operators inside in person. The acoustic optimisation aspect is particularly relevant in cases where internal operators are present, ensuring a more comfortable working environment while maintaining optimal system performance. These efforts aim to reduce consumption, enhance sustainability, and ensure a more efficient operating environment for IT equipment. Figure 5 gives an overview of variables, models and use cases of the data centre DT.

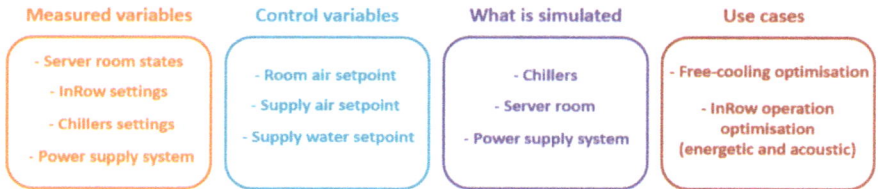

Fig. 5. Overview of variables, models and use cases of the data centre DT.

The creation of a data centre DT could help to monitor room temperatures and cooling supply, prevent failures, and suggest efficient control strategies. As a result, it is expected to improve the Power Usage Effectiveness (PUE) by optimising room temperature settings and reducing cooling supply, leading to lower electricity expenses and CO_2 savings.

The Z3 cooling plant operation will be monitored as well, ensuring efficient performance, reducing failures, and improving control strategies. This approach is anticipated to lower the PUE by enhancing the plant overall efficiency, thereby cutting electricity costs, and contributing to CO_2 savings. Monitoring efforts will focus on the chiller operation mode – whether it is working in compression cooling, partial free cooling, or free cooling – along with the electricity consumption. Currently, free-cooling is activated only when the external temperature falls below 10 °C, which limits its utilisation. By integrating SIMBOTS for real-time simulation within the Software-in-the-Loop (SiL) environment, more advanced dynamic control strategies can be implemented. This approach enables the extension of free cooling operation across a broader range of conditions, enhancing its overall contribution to energy savings.

4.2 Active Heat Recovery Digital Twin

The heat recovery DT should be designed to monitor, control, and simulate the operation of the WW-HP integrated to BL26 existing heating system. To ensure efficient operation, several key variables are measured, including the electricity consumption of the WW-HP, water temperatures, and water flow rates. These parameters provide essential insights into system performance and energy usage. Control strategies focus on adjusting the WW-HP temperature setpoints and operation modes, as well as regulating flow rates or temperature differences (ΔT). Simulations could focus on a physics-based and/or data-driven BL26 model to estimate the heating needs and on a WW-HP model to study the system behaviour under different conditions.

The main use cases include integrating the WW-HP into the existing heating plant, optimising system management during mid-seasons, and implementing free cooling strategies to enhance energy efficiency. A key challenge is the non-trivial coupling of the WW-HP with the existing system, requiring careful analysis of energy flows. During mid-seasons, when the data centre needs to dissipate more energy than the BL26 requires, the WW-HP will operate alongside the Z3 chillers to balance the load. In winter conditions, the WW-HP will work in synergy with the free cooling mode of the Z3 chillers, further improving overall efficiency. Figure 6 gives an overview of variables, models and use cases of the heat recovery DT.

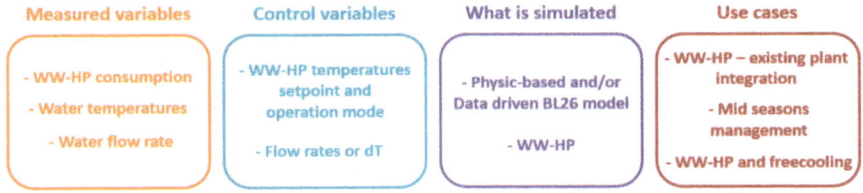

Fig. 6. Overview of variables, models and use cases of the heat recovery DT.

The water-water heat pump extracts waste heat from the server room and integrates it into the existing gas boiler heating system of building BL26. The main objective of the heat recovery DT is to verify the correct integration of the water-to-water heat exchange system, ensuring that the control strategy prioritizes heat pump operation. At the same time, it is essential to monitor the electricity consumption of both the heat pumps and the circulation pumps, as well as to evaluate the thermal energy effectively produced. Another key aspect is the analysis of the building heating needs and the performance of the heat pump, with the goal of optimising system efficiency and minimizing overall energy costs. To achieve this, it is necessary to track key parameters such as supply and return temperatures, water flow rates at the evaporator and condenser, and water temperatures at various points in the system. Finally, based on the collected data, the system should be able to suggest optimal control variable trajectories, thereby maximizing overall efficiency.

5   Conclusions

The paper presents a brief introduction to HYCOOL-IT project, with a description of data centre Z3 and its WHR system. The implementation of a WHR system in the Z3 data centre, using a WW-HP connected to the heating system of Building BL26, will enable a significant reduction in natural gas consumption, energy costs, and CO_2 emissions, while simultaneously improving the overall energy efficiency of the campus. To ensure correct management and optimisation of the system, a DT environment is currently under development. This article outlines its general architecture, emphasising the crucial role of integrating physical devices with cloud-based analytics for real-time monitoring and optimisation. At the core of this process is the IoT HUB service, which facilitates efficient data collection, transmission, and processing, enabling smarter and more proactive system management. By combining real-time monitoring, simulation models, and predictive control, the DT continuously optimises data centre operations. Leveraging predictive modelling and optimisation techniques, it enhances energy efficiency, stability, and the overall reliability of critical data centre infrastructure. The creation of two DTs – the data centre DT and the Active (WW-HP) Heat Recovery DT – could be an effective approach for managing energy use, controlling key variables, and testing different management strategies without disrupting normal operations. By continuously monitoring and adjusting parameters such as temperature setpoints, cooling settings, and flow rates, the DTs will help to enhance overall performance, improve PUE, and reduce energy costs and CO_2 emissions. Looking ahead, the integration of the WW-HP system with the BL26 existing heating system represents an innovative approach to coupling data centre operations with building energy needs. The simulations and real-time data analysis will enable further optimisation, particularly during mid-seasons and winter when energy efficiency is critical. The expected outcomes will not only contribute to lowering operational costs but will also provide valuable insights into the broader application of DT technology in industrial settings, paving the way for more sustainable and efficient energy management practices across similar infrastructures. As the project progresses and more mature results become available, original findings and novel contributions are expected to be presented in future publications.

Acknowledgements. HYCOOL-IT project is funded by the European Union's Horizon Europe program under Grant Agreement No. 101138623 and is scheduled to run from December 1, 2023, to November 30, 2026.

References

1. Grieves, M.: Origins of the Digital Twin Concept (2016). https://doi.org/10.13140/RG.2.2. 26367.61609
2. Opoku, D.G.J., Perera, S., Osei-Kyei, R., Rashidi, M.: Digital twin application in the construction industry: a literature review. J. Build. Eng. **40**, 102726 (2021). https://doi.org/10. 1016/J.JOBE.2021.102726
3. Abdelrahman, M., Macatulad, E., Lei, B., Quintana, M., Miller, C., Biljecki, F.: What is a digital twin anyway? Deriving the definition for the built environment from over 15,000

scientific publications. Build. Environ. **274**, 112748 (2025). https://doi.org/10.1016/J.BUI LDENV.2025.112748

4. Ozturk, G.B.: Digital twin research in the AECO-FM industry. J. Build. Eng. **40**, 102730 (2021). https://doi.org/10.1016/j.jobe.2021.102730

5. Boje, C., Guerriero, A., Kubicki, S., Rezgui, Y.: Towards a semantic construction digital twin: directions for future research. Autom. Constr. **114**, 103179 (2020). https://doi.org/10.1016/J. AUTCON.2020.103179

6. Sphere BIM Digital Twin Platform - White Paper Q4: Digital Twin Definitions for Buildings - How to Move Digital Twin Environments to the AECOO Sector (2019). [Online]. www.sph ere-project.eu

7. Singh, M., Fuenmayor, E., Hinchy, E.P., Qiao, Y., Murray, N., Devine, D.: Digital Twin: Origin to Future. DPI AG (2021). https://doi.org/10.3390/asi4020036

8. AlBalkhy, W., Karmaoui, D., Ducoulombier, L., Lafhaj, Z., Linner, T.: Digital twins in the built environment: definition, applications, and challenges. Autom. Constr. **162**, 105368 (2024). https://doi.org/10.1016/J.AUTCON.2024.105368

9. European Committee for Standardization: DRAFT prEN 18162 Building Information Modelling (BIM) - Digital twins applied to the built environment - Concept and definitions (2025) [Online]. www.bsigroup.com

10. Bortolini, R., Rodrigues, R., Alavi, H., Vecchia, L.F.D., Forcada, N.: Digital Twins' Applications for Building Energy Efficiency: A Review. MDPI (2022). https://doi.org/10.3390/en1 5197002

11. Liu, Z., Li, M., Ji, W.: Development and application of a digital twin model for Net zero energy building operation and maintenance utilizing BIM-IoT integration. Energy Build **328**, 115170 (2025). https://doi.org/10.1016/J.ENBUILD.2024.115170

12. Lu, Q., Xie, X., Parlikad, A.K., Schooling, J.M.: Digital twin-enabled anomaly detection for built asset monitoring in operation and maintenance. Autom. Constr. **118**, 103277 (2020). https://doi.org/10.1016/J.AUTCON.2020.103277

13. Zhao, J., Feng, H., Chen, Q., Garcia de Soto, B.: Developing a conceptual framework for the application of digital twin technologies to revamp building operation and maintenance processes. J. Build. Eng. **49**, 104028 (2022). https://doi.org/10.1016/j.jobe.2022.104028

14. Elshabshiri, A., Ghanim, A., Hussien, A., Maksoud, A., Mushtaha, E.: Integration of building information modeling and digital twins in the operation and maintenance of a building lifecycle: a bibliometric analysis review. J. Build. Eng. **99**, 111541 (2025). https://doi.org/10. 1016/J.JOBE.2024.111541

15. Cespedes-Cubides, A.S., Jradi, M.: A Review of Building Digital Twins to Improve Energy Efficiency in the Building Operational Stage. Springer Nature (2024). https://doi.org/10.1186/ s42162-024-00313-7

16. Seghezzi, E., et al.: Towards an occupancy-oriented digital twin for facility management: test campaign and sensors assessment. Appl. Sci. (Switz.) (2021). https://doi.org/10.3390/app110 73108

17. Arowoiya, V.A., Moehler, R.C., Fang, Y.: Digital twin technology for thermal comfort and energy efficiency in buildings: a state-of-the-art and future directions. Energy Built Envir. **5**(5), 641–656 (2024). https://doi.org/10.1016/J.ENBENV.2023.05.004

18. Yuan, X., Liang, Y., Hu, X., Xu, Y., Chen, Y., Kosonen, R.: Waste heat recoveries in data centers: a review. Renew. Sustain. Energy Rev. **188**, 113777 (2023). https://doi.org/10.1016/ J.RSER.2023.113777

19. IDP Ingenieria y Arquitectura Iberia SL. https://www.idp.es/en/idp-digital/. last accessed 08 Apr 2025

20. Functional Mock-up Interface (FMI). https://fmi-standard.org/. last accessed 08 Apr 2025

21. Romagnosi, M., Aprile, M., Dénarié, A.: Energy and economic simulation of a renewable energy community applied to a new generation ultra-low temperature district heating and cooling network. In: E3S Web of Conferences. EDP Sciences (2024). https://doi.org/10.1051/e3sconf/202452305003

22. Famiglietti, J., Aprile, M., Spirito, G., Motta, M.: Net-Zero climate emissions districts: potentials and constraints for social housing in Milan. Energies (Basel) (2023). https://doi.org/10.3390/en16031504

23. ISO 52016-1:2017 - Energy performance of buildings - Energy needs for heating and cooling, internal temperatures and sensible and latent heat loads - Part 1: Calculation procedures, ISO 52016-1:2017 (2017)

24. EcoStruxure Platform by Schneider Electric. https://www.se.com/it/it/work/campaign/innovation/platform.jsp. last accessed 08 Apr 2025

Digital Twin for Datacenter: HPC4AI UniTO Case Study

Viviana Vaccaro[1] (ID), Robert Birke[2] (ID), Silvia Meschini[2(✉)] (ID),
Lavinia Chiara Tagliabue[2] (ID), Sergio Rabellino[2] (ID), Pablo Vicente Legazpi[3] (ID),
and Marco Andinucci[2] (ID)

[1] Politecnico Di Milano, Milan, Italy
[2] University of Turin, Turin, Italy
silvia.meschini@unito.it
[3] BDTA, Madrid, Spain

Abstract. The HPC4AI (High-Performance Computing for Artificial Intelligence) datacenter at the University of Turin's Computer Science Department was established to meet the rapidly growing computational demands of interdisciplinary AI research. HPC4AI innovates by redefining the traditional roles of Cloud and High-Performance Computing (HPC) systems, where the Cloud provides a modern interface for HPC, and HPC acts as an accelerator for Cloud applications. To date, it has supported over 40 research projects spanning diverse fields such as astronomy, medicine, and human sciences. Additionally, HPC4AI serves as a research and development platform for exploring, developing, and testing novel datacenter technologies. It features a variety of experimental computing platforms and the first prototype of a two-phase evaporative server cooling system. This work outlines the operational management of HPC4AI, highlighting challenges, lessons learned, and key opportunities related to digital twins for datacenters.

Keywords: HPC4AI · Datacenter · Digital Twin · Artificial Intelligence · High Performance Computing (HPC) · Energy modeling

1 Introduction

The HPC4AI datacenter was established to meet the evolving computational demands of Artificial Intelligence (AI) applications by rethinking the conventional separation between Cloud and HPC systems [1]. Rather than treating them as distinct paradigms, HPC4AI integrates Cloud and HPC into a symbiotic system, in which the Cloud provides flexible, on-demand services, such as model inferences (e.g., the PRAISE cardiopathic risk score) via virtual machines and containers. Meanwhile, compute intensive tasks like model training are delegated to the HPC clusters as batch jobs. This architecture enables the Cloud to serve as a modern, user-friendly interface for HPC, while HPC accelerates performance-critical components of Cloud applications. A practical illustration of this integration is represented by the Jupyter Workflow. Jupyter notebooks offer a feature-rich, cell-based web interface that seamlessly integrates code snippets with descriptive

A. Jurelionis et al. (Eds.): BDTIC 2025, LNCE 775, pp. 99–110, 2026.
https://doi.org/10.1007/978-3-032-09040-9_9

metadata [2, 3], facilitating remote execution and making them popular among domain experts and AI practitioners. Unlike traditional Jupyter notebooks, which execute code cells sequentially on a single server, Jupyter Workflow enables users to orchestrate complex workflows and execute them across heterogeneous hybrid HPC-Cloud infrastructures. In this configuration, the Cloud hosts the web-based frontend, while distributed execution is coordinated and offloaded to the HPC backend. This model supports scalability, reproducibility, and resource optimisation, and aligns with emerging concepts of Digital Twins (DTs) and scientific workflows in AI-driven research environments [5, 7, 12]. Indeed, DTs represent a transformative innovation for data center management, enabling real-time monitoring, predictive maintenance, and energy optimization. By virtually replicating physical infrastructure, DTs support rapid intervention in case of anomalies, can improve system resilience and reduce environmental impact, key features to shift toward more efficient and sustainable HPC environments. Their integration with IoT and AI further enhances dynamic system control, supporting predictive analytics and adaptive cooling strategies [6, 8, 9].

HPC4AI operates as a federated competence center with two main hubs: one located at the University of Turin's Computer Science Department and the other at Politecnico di Torino. The remainder of this paper focuses specifically on the University of Turin (UNITO) hub (Fig. 1), which serves as a case study for exploring challenges, current gaps and opportunities in the development of DTs for datacenters.

Fig. 1. HPC4AI@UNITO datacenter

2   Design of HPC4AI at the Computer Science Department, UNITO

HPC4AI is hosted in a custom-designed 250 kVA data center located at the University of Turin's Computer Science Department. The facility is engineered to meet Tier-III availability standards, ensuring an uptime of 99.982%, equivalent to no more than 1.6

h of annual downtime [7]. Redundant power and cooling systems guarantee continuous operation. In the event of short power interruptions, two UPS units temporarily power the IT equipment until a diesel generator can restore power to both IT and cooling systems. Over four years of operation, the datacenter experienced only one downtime, necessitated by the connection of a second chiller to the UPS-protected power supply.

Power Distribution: Power is distributed through ceiling-mounted crossbars, allowing for maximum flexibility in connecting rack PDUs. The datacenter comprises 16 racks, each equipped with two PDUs that monitor load at the phase level. Servers are connected to balance the load across the three phases, while the Building Management System (BMS) continuously monitors power at the distribution transformer for real-time Power Usage Effectiveness (PUE) monitoring.

Cooling System: The datacenter employs a cold-hot aisle design, with the hot aisle centrally located and eight racks on either side, topped by an extractor hood. Cold air is supplied from above, leveraging natural air circulation to reduce fan energy consumption. Two adiabatic chillers on the roof operate in three modes: free cooling, evaporation, or compression, depending on weather conditions. These chillers primarily function in free or evaporative modes, enhancing energy efficiency due to their strategic location in a windy area with dry air from surrounding mountains. Water consumption is notably low, with only 1671 cubic meters used over four years, equivalent to just 5.35 times the average per capita water consumption in Italy (based on 2021 data) [4].

Showroom and Educational Value: Glass walls on two sides of the server room provide students and visitors with a clear view, significantly enhancing the educational value of the datacenter. A challenge was soundproofing the facility, given its proximity to departmental offices. The use of triple glazing reduced noise levels by up to 92 dB, effectively silencing high-speed cooling fans. Additionally, cooling pipes on the exterior were constructed from specialized soundproofing materials to minimize noise pollution both inside and outside the facility.

2.1 A Multitenant Cloud

The OpenStack-based cloud system features a robust infrastructure, comprising over 2400 physical cores, 60 TB of RAM, and 120 GPUs spanning three generations (NVIDIA T4, V100, and A40). The system is interconnected via a high-speed 25 Gb/s network and offers four distinct storage classes, each with unique characteristics and cost structures. These storage options range from ultra-fast Ceph storage utilizing nVMe disks (40 TB) to fast SSD storage based on Dell EMC Unity (1 PB) and extend to spinning disk-based storage for cold data (1 PB), which is backed up using Dell EMC Avamar (1 PB). Additionally, a smaller 4-node system replicates the cloud infrastructure for testing and research purposes. Users can request virtual machines (VMs) and utilize UNITO's custom management tools to deploy private, secure, and elastic instances of Kubernetes. These Kubernetes instances are enhanced with single sign-on capabilities through Keycloak, providing robust identity and access management features.

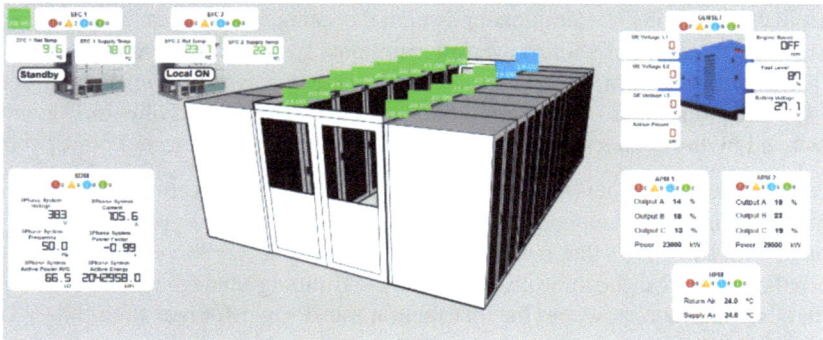

Fig. 2. HPC4AI BMS system overview and data display.

2.2 Modular HPC System

The HPC cluster comprises heterogeneous compute nodes:

- 68 Intel nodes (2x Xeon CPU E5–2697 v4, 36 cores with HT disabled, 128 GB RAM, OmniPath 100 Gb/s)
- 4 Arm nodes (Ampere Altra 80 cores, 512 GB RAM, 2xA100 GPUs, 2xBlueField2 DPUs, Ethernet 100 Gb/s)
- 4 Intel nodes (2x Xeon Gold 6230, 40 cores with HT disabled, 1 TB RAM, 1xT4 + 1xV100 GPUs, Ethernet 100 Gb/s)
- 4 Nvidia nodes (1x Grace CPU, 72 cores, 573 GB RAM, 1xH100 GPU, Ethernet 100 Gb/s) network.

These are supported by 2 all-flash HPC storage systems: BeeGFS hosting user homes (50 TB) and LUSTRE as scratch partition (20 TB). The system is managed by UrgentSLUM, a self-developed SLURM extension that manages the booking of nodes via a web calendar.

2.3 Experimental System

HPC4AI serves as a platform for hosting experimental and prototype architecture, including:

- SiFive and Milk-V RISC-V servers,
- Esperanto RISC-V boards
- NVIDIA development kits equipped with Bluefield DPUs

These systems support cutting-edge research in innovative computing architectures. Notably, HPC4AI is home to the first commercial prototype of a two-phase evaporative cooling system designed for high-power density GPU-enabled servers.

This prototype features a dual-socket Intel Xeon Platinum 8458P configuration paired with four NVIDIA H100 SXM2 GPUs. The evaporator is engineered to dissipate up to 1000 W per socket, handling thermal densities exceeding 70 W/cm^2. Two-phase cooling is anticipated to surpass the efficiency of current liquid cooling solutions, offering additional benefits such as the use of dielectric (environmentally friendly and non-toxic) gas. This design mitigates the risks associated with liquid cooling, such as damage from pipe ruptures. Furthermore, the system utilizes thin, low-pressure pipes, reducing costs and enhancing overall efficiency by leveraging evaporation for heat dissipation.

3 Monitoring System for the Digital Infrastructure

Comprehensive monitoring is essential for ensuring smooth operation and effective management of the HPC4AI datacenter. To achieve this, three overarching monitoring systems are employed, each targeting different aspects of data center operations: one focused on DC servers, another integrated into the Building Management System (BMS), and a standalone system uploading data to the cloud with complementary sensors.

Power and Cooling: The Vertiv Critical Insight BMS plays a crucial role in controlling and monitoring the status of power distribution and cooling equipment. The main dashboard provides a concise summary of the system's health status (Fig. 2). In the event of anomalies or faults, alarms are triggered to facilitate prompt interventions. For power management, the system tracks energy efficiency to identify opportunities for improvement, using sensors from the distribution transformer down to the PDUs in the racks. For cooling, it monitors air temperature and humidity at both the chiller and rack levels, ensuring balanced cooling and preventing hot spots. Additionally, it surveils the dew point to prevent damage from water condensation on IT equipment.

Air Quality: To ensure the longevity of IT equipment, monitoring extends beyond temperature and humidity to include various particulate matter in the air. Three LaCentralina air quality monitoring stations [5] are used to track pollutants such as particle matters of different sizes (PM1, PM2.5, PM4, and PM10) and gases (CO_2, tVOC, O_3, NH_3, CO, NO_2, and CH_2O). Particulate matter can be detrimental to IT equipment by depositing dust on components, potentially clogging airflows and fans.

IT Systems: The operational health of all IT subsystems, OpenStack, SLURM, Ceph, and Unity, is monitored through their respective dashboards. In addition, the Pandora FMS monitoring platform is employed to track real-time resource usage (CPU, RAM) and to detect potential performance bottlenecks across the infrastructure.

Power Usage Effectiveness (PUE) [6]: Based on monitored metrics, particular attention is given to tracking the Power Usage Effectiveness (PUE) of the data center. PUE quantifies how efficiently a data center uses energy by comparing IT equipment energy usage to total energy consumption. A PUE value closer to 1 indicates ideal operational efficiency, where most energy is used for IT resources rather than cooling or other purposes. The recent historical record of energy consumption and the achieved shows a PUE of 1.1, notably lower than the estimated global average PUE of 1.56 for data centers [6, 12].

4 Datacenter Modeling for Digital Twin

The data coming from the sensors and the information related to the data center are converging in a BIM model that will be used as a basis for the DT of the HPC4AI in the modelling phase [7, 11] and later as a support to visualize the data (Fig. 3). The BIM model was based on CAD files and used for building energy modeling.

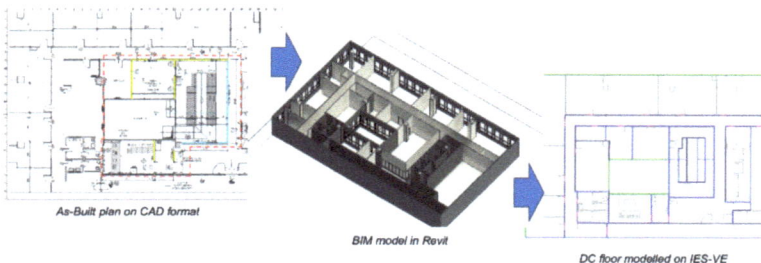

As-Built plan on CAD format

BIM model in Revit

DC floor modelled on IES-VE

Fig. 3. Model creation to simulate and visualize the DC indoor conditions and create the DT

The primary objective of developing a DT for the data center is to consolidate critical operational data and system insights into a virtual replica, enabling real-time lifecycle management and dynamic performance simulations. DTs represent a transformative paradigm in infrastructure management by supporting predictive maintenance, anomaly detection, and scenario-based optimization of both energy use and equipment operation [9, 12]. When integrated with IoT sensor networks, AI-driven analytics, and BIM, DTs provide a robust framework for simulating thermal and electrical behavior, assessing system resilience, and improving power usage effectiveness (PUE) [11]. BIM accelerates DT deployment by supplying precise spatial and semantic modeling of architectural and mechanical systems, and enhances the DT's reliability in energy optimization, fault detection, and long-term sustainability planning [8–11].

The DT for the HPC4AI data center is being developed to progressively incorporate these capabilities with the goal of enhancing energy efficiency and operational reliability in high-performance computing environments. The initial phase focused on constructing a validated, simulation-ready model to evaluate energy performance, indoor environmental quality, and thermal exchanges between the data center and its host facility under various operational conditions. This foundational model supports the integration of future DT features such as real-time monitoring, data-driven control strategies, and adaptive system optimization. Key priorities include managing internal heat generation, maintaining optimal environmental parameters for computing equipment, and minimizing energy losses through efficient heat dissipation. A dynamic simulation model has been established as a preliminary implementation, structured around the detailed configuration of system parameters and the evolving DT framework.

Modeling Approach: Dynamic simulations were conducted using IES Virtual Environment (IES VE), a leading building performance simulation platform (Fig. 4).

Climate Data Integration: Local environmental conditions critically influence free and evaporative cooling efficiency. Climate files integrated into the simulation software were selected based on the site's geographic location. These files provide granular datasets, including outdoor temperature, relative humidity, wind speed/direction, and solar radiation, ensuring accurate representation of regional climate patterns and reliable energy performance assessments.

Thermal Zone Definition: Geometric modeling formed the foundation for accurately replicating the data center's layout, ensuring fidelity in representing spatial configurations, components, and thermal zone interactions. The model was cross verified against technical drawings and physical measurements to ensure dimensional accuracy, volumetric consistency, and zone adjacency relationships. Four distinct thermal zones were defined: the server room, cold aisle, hot aisle, and supply/return air plenums.

The process encompassed rigorous validation of geometric parameters, thermal zone boundaries, and system interactions to align the digital model with real-world operational dynamics.

4.1 Geometric Modeling

Thermal zones within the data center were delineated based on their specific functions to ensure precise control over airflow and thermal conditions. The zones were categorized into primary, auxiliary, and adjacent spaces.

Primary Spaces: *Server Room/Cold Aisle:* This area is responsible for distributing cooled air to racks via strategically placed supply grilles. *Hot Aisle:* An isolated area where hot air from racks accumulates, designed for efficient air recovery through the return plenum. *Supply and Return Plenums:* Elevated spaces facilitating cooled air distribution and hot air recovery.
Auxiliary Spaces: *Control Room:* Dedicated for monitoring and managing operations. *Water Treatment Room:* Supports evaporative cooling systems. *Q.E. and UPS Room:* Houses electrical panels and uninterruptible power supplies.
Adjacent Spaces: *Generic Rooms, Corridors, WCs:* Neighboring spaces influencing thermal exchanges without being part of the main cooling system.

To enhance simulation accuracy, the supply and return plenums, racks, and hot aisles were modeled as separate geometric zones. This allowed for detailed analysis of temperature, pressure, and airflow velocity differences within each zone. Construction materials were characterized using the IES VE database, supplemented by project data, to ensure realistic thermal modeling. Each element was defined based on its physical and thermal properties, such as the Hot Aisle Containment System, which minimizes hot and cold air mixing to maximize heat recovery efficiency.

4.2 Use Profiles

Project profiles are essential tools for defining operational schedules and thermal settings for HVAC systems and internal loads across various zones. These profiles, configured within the APpro module, enable precise control over system on/off schedules, temperature setpoints, setback values, and operational patterns. By tailoring these profiles to match specific space needs, the simulation accurately reflects energy consumption and environmental conditions.

Adjacent Spaces (Offices and Auxiliary Rooms):

Operating Hours: HVAC systems operate from 8:00 AM to 6:00 PM, with setbacks during off-peak hours.
Temperature Setpoints and Setbacks:

Heating: 20 °C during operation, 15 °C during unoccupied periods.
Cooling: 24 °C during operation, 30 °C during unoccupied periods.
Load Profiles: Internal gains from lighting, equipment, and occupancy align with typical office schedules, reflecting realistic energy use patterns.

Server Room:

Temperature Control: Maintained within the ASHRAE-recommended range of 18 °C to 27 °C to ensure equipment reliability and flexibility for cooling capacities.
Humidity Control: Relative humidity is kept between 20% and 80%, with continuous HVAC operation to prevent hotspots and maintain uniform conditions.
IT Equipment Load: Modeled as a constant heat source using measured power consumption data to capture steady-state thermal behavior accurately.

By implementing these tailored profiles, the model accurately reflects distinct operational patterns and thermal requirements of each zone. For adjacent spaces, energy consumption is minimized during unoccupied hours, aligning with efficiency practices. In the server room, continuous HVAC operation ensures critical temperature and humidity thresholds are maintained, safeguarding IT equipment functionality. Figure 4 illustrates a temperature simulation over a year, demonstrating the robust foundation for simulating energy performance and environmental conditions.

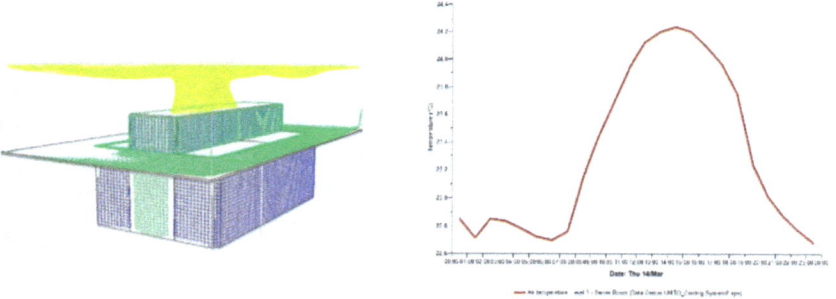

Fig. 4. Computational Fluid Dynamics (CFD) Hot Aisle Containment Temperature Distribution, March typical day trend of temperature in the Server Room

4.3 Thermal Modeling

The configuration of thermal templates in IES VE was crucial for accurately capturing the distinct thermal characteristics of each zone within the data center. By assigning standardized thermal properties across zones, thermal templates enable consistent and flexible management of parameters, ensuring that the model accurately reflects variations in thermal behavior while maintaining computational efficiency. Each zone was specifically configured to reflect its unique thermal attributes, including internal heat gains, airflow rates, and boundary conditions. Real-time operational data from sensors installed throughout the data center provided a robust basis for calibrating the model.

This data included power consumption, rack-level temperatures, and humidity levels, allowing the model to closely align with observed conditions and enhance its predictive accuracy.

4.4 Integration of Electrical Systems

Linked to thermal management are electrical systems. The final PuE KPI is directly connected with electrical performance. So thermal management is helping to describe the losses of the system. Due to the sampling time frame (5 or 10 min), electrical representation of components must be done following a phasor representation (or DQ representation). A common application is in the steady-state analysis of an electrical network powered by time varying current where all signals are assumed to be sinusoidal with a common frequency. Phasor representation allows the analyst to represent the amplitude and phase of the signal using a single complex number. A simple simulation model for real time can be implemented, using some of the monitored variables as input and calculating the rest of monitoring pack. The comparison of these values in each time step can provide information of consistency of power meters values and a nominal exploitation of the data center, detecting any deviation. This logic is reflected in the electrical model illustrated in Fig. 5, which shows the representation of the data center's power feed and its integration into the simulation workflow.

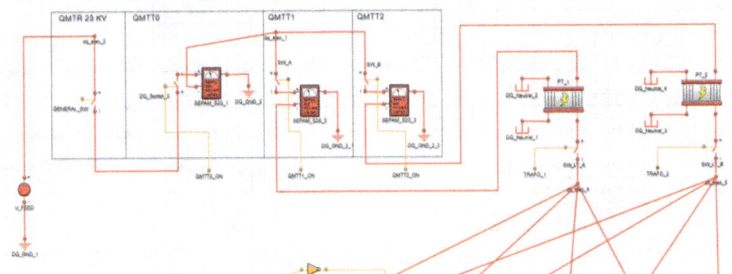

Fig. 5. Electrical model of data center feed.

In a similar way, integration of other systems is possible. For example, new cooling machines of small components of the data center. These partial systems can be modelled independently and can be run in separate services. Integration times must be inside the sampling time frame and result may be integrated with monitoring signals, whether as redundant information to detect faults or as extra monitoring or KPIs information. An example of this modular modeling is provided in Fig. 6, which shows the chiller system representation integrated into the DT for detailed thermal behavior simulation.

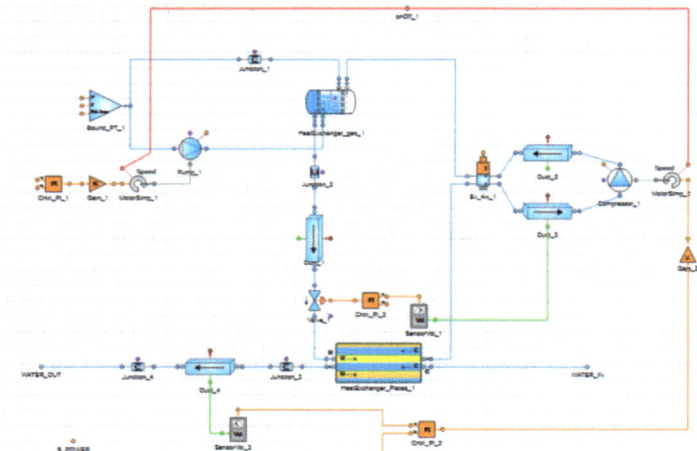

Fig. 6. Chiller model

5   Discussion and Conclusions

The development of a digital twin for the High-Performance Computing for Artificial Intelligence (HPC4AI) facility at the University of Turin represents a groundbreaking approach to energy management in data centers. This initiative involves integrating a sophisticated monitoring system with a digital model that enables energy simulation and validation against real-world data. By capturing critical performance metrics, the system facilitates real-time analysis and decision-making, ensuring the reliability of the digital model through iterative validation. This process enhances operational efficiency and lays the foundation for comprehensive lifecycle management of the data center. The digital twin will enable proactive identification of inefficiencies, predictive maintenance, and strategic planning for future expansions, thereby reducing operational costs and environmental impact. By simulating various scenarios, stakeholders can assess the implications of design choices on energy consumption and performance, aligning operations with sustainability goals. This approach marks a significant shift from reactive to proactive energy management, addressing potential issues before they escalate. As the digital twin is refined, it is expected to become a valuable tool for optimizing energy use throughout the data center's lifecycle, contributing to more sustainable practices in high-performance computing environments.

Acknowledgment. This work has been supported by the Spoke "FutureHPC & BigData" of the ICSC-Centro Nazionale di Ricerca in "High Performance Computing, Big Data and Quantum Computing", funded by the European Union - NextGenerationEU and by the DYMAN project funded by the European Union - European Innovation Council under G.A. n. 101161930.

References

1. F. D'Ascenzo, O. D. Filippo, G. Gallone, G. Mittone, M. A. Deriu, M. Iannaccone, A. Ariza-Sole´, C. Liebetrau, S. Manzano-Ferna´ndez, G. Quadri, T. Kinnaird, G. Campo, J. P. S. Henriques, J. M. Hughes, A. Dominguez-Rodriguez, M. Aldinucci, U. Morbiducci, G. Patti, S. Raposeiras-Roubin, E. Abu-Assi, G. M. D. Ferrari, F. Piroli, A. Sagli- etto, F. Conrotto, P. Omede´, A. Montefusco, M. Pennone, F. Bruno, P. P. Bocchino, G. Boccuzzi, E. Cerrato, F. Varbella, M. Sperti, S. B. Wilton, L. Velicki, I. Xanthopoulou, A. Cequier, A. Iniguez-Romo, I. M. Pousa, M. C. Fernandez, B. C. Queija, R. Cobas-Paz, A. Lopez-Cuenca, A. Garay, P. F. Blanco, A. Rognoni, G. B. Zoccai, S. Biscaglia, I. Nunez- Gil, T. Fujii, A. Durante, X. Song, T. Kawaji, D. Alexopoulos, Z. Huczek, J. R. G. Juanatey, S.-P. Nie, M. aki Kawashiri, I. Colonnelli, B. Cantalupo, R. Esposito, S. Leonardi, W. G. Marra, A. Chieffo, U. Michelucci, D. Piga, M. Malavolta, S. Gili, M. Mennuni, C. Montalto, L. O. Visconti, and Y. Arfat, "Machine learning-based prediction of adverse events following an acute coronary syndrome (praise): a modelling study of pooled datasets," The Lancet, vol. 397, no. 10270, p. 199–207, 2021
2. Colonnelli, I., et al.: Distributed workflows with jupyter. Future Gener. Comput. Syst. **128**, 282–298 (2022)
3. Kluyver, T., et al.: Jupyter notebooks – a publishing format for reproducible computational workflows. In: Loizides, F., Schmidt, B. (eds.) Positioning and Power in Academic Publishing: Players, Agents and Agendas, pp. 87–90. IOS Press (2016)
4. TISTAT: Le Statistiche dell'ISTAT Sull'acqua (2024). https://www.istat.it/wpcontent/uploads/2024/03/Report-GMA-Anno-2024.pdf
5. a-GRISU: Lacentralina 5–22. https://www.agrisu.com/products/lacentralina-5-22/
6. Institute, U.: Uptime Institute Global Data Center Survey 2024. https://datacenter.uptimeinstitute.com/rs/711-RIA145/images/2024.GlobalDataCenterSurvey.Report.pdf
7. IES: Virtual Environment. https://www.iesve.com/software/virtualenvironment
8. Rasheed, A., San, O., Kvamsdal, T.: Digital twin: values, challenges and enablers from a modeling perspective. IEEE Access **8**, 21980–22012 (2020). https://doi.org/10.1109/ACCESS.2020.2970143
9. Zhang, Y., Yu, Z., Zeng, D.: A digital twin approach for predictive maintenance of data center cooling systems. Energy Build. **231**, 110591 (2021)
10. Qi, Q., Tao, F.: Digital twin and big data towards smart manufacturing and industry 4.0: 360 degree comparison. IEEE Access **6**, 3585–3593 (2018). https://doi.org/10.1109/ACCESS.2018.2793265
11. Pinmas, P., et al.: Towards digital twin data center using building information modeling and real-time data sensing. In: 2022 37th International Technical Conference on Circuits/Systems, Computers and Communications (ITC-CSCC). IEEE (2022)
12. Zohdi, T.I.: A digital-twin and machine-learning framework for precise heat and energy management of data-centers. Comput. Mech. **69**(6), 1501–1516 (2022)

Digital Tools for Sustainable Renovation of Heritage Buildings: From HBIM to Low-Carbon Material Selection

Erika Svytytė, Vytautas Bocullo⬤, and Lina Seduikytė(✉) ⬤

Kaunas University of Technology, Studentu Str. 48, 51367 Kaunas, Lithuania
lina.seduikyte@ktu.lt

Abstract. This paper presents a case study on applying digital tools for the sustainable renovation of a cultural heritage building, the Mikas and Kipras Petrauskas House in Kaunas, Lithuania. The study integrates Heritage Building Information Modeling (HBIM), indoor environmental quality (IEQ) monitoring, and life cycle assessment (LCA) to evaluate renovation solutions in alignment with smart and sustainable city goals. A high-accuracy HBIM model was developed using photogrammetry and laser scanning, achieving less than 2% deviation from validated geometric dimensions. The model provided a reliable foundation for analytical workflows and demonstrated how digital capture can reduce time and mitigate reliance on outdated documentation. Indoor climate measurements indicated critical humidity conditions during the cold season, highlighting the need for improved environmental controls to preserve both occupant comfort and material preservation. LCA was employed to assess the embodied carbon of renovation materials, and alternative selections led to a 22% reduction in CO_2e emissions. The results confirm that integrating HBIM, IAQ analysis, and LCA provides a robust methodology for data-driven decision-making in heritage renovation. The approach also supports the future development of digital twin frameworks that balance environmental sustainability with cultural value.

Keywords: Heritage Building · HBIM · IAQ · LCA · Renovation · Sustainability

1 Introduction

The renovation of cultural heritage buildings is becoming increasingly important in the context of sustainable urban development. In the European Union, buildings account for approximately 40% of total energy consumption, and more than 80% of existing buildings were constructed before 1990. Many of these are heritage buildings, which often remain unrenovated and inefficient in terms of energy and environmental performance [1, 2]. In Lithuania alone, there are more than 9,000 registered cultural heritage buildings, over 5,000 of which are structures such as houses, churches, schools, and museums that are of significant historical and architectural value [3].

Modernizing and preserving these structures must address sustainability and conservation objectives. The integration of digital tools, particularly Building Information

© The Author(s) 2026
A. Jurelionis et al. (Eds.): BDTIC 2025, LNCE 775, pp. 111–122, 2026.
https://doi.org/10.1007/978-3-032-09040-9_10

Modeling (BIM) and its specialized form for heritage assets Heritage Building Information Modeling (HBIM), has created new opportunities for informed decision-making in renovation processes [4–7]. HBIM enables the accurate reconstruction of building geometry using techniques such as photogrammetry and laser scanning, and it serves as a foundation for further simulation and analysis, including environmental performance modelling and life cycle assessment [8, 9].

Beyond structural and energy aspects, heritage buildings' indoor environmental quality (IEQ) is receiving growing attention, particularly concerning thermal comfort, humidity, and air pollutant levels. Poor indoor air quality (IAQ) affects occupant well-being and accelerates the degradation of historical materials. Digital simulations and sensor integration can help optimize IAQ during both the design and operational phases, especially in buildings where physical interventions are limited by preservation rules [11].

Despite these advancements, HBIM is still predominantly used for geometric modelling and documentation, with limited applications in operational performance assessment. Most studies focus on structural visualization and maintenance planning rather than energy or indoor environmental quality (IEQ) analysis [5, 10]. Furthermore, the current standards for HBIM models, such as Level of Detail (LOD), do not fully address the data needs for precise microclimate or energy simulations, limiting the reliability of predictive analyses compared to real-world measurements [12, 13].

Meanwhile, life cycle assessment (LCA) provides a critical method for quantifying the environmental impact of renovation choices. In heritage contexts, LCA supports material and system selection that reduces embodied carbon while maintaining architectural authenticity. Integrating LCA with HBIM and performance modelling tools can yield renovation strategies that are both historically sensitive and environmentally optimized [14].

The concept of digital twins virtual representations of physical assets that integrate real-time data and simulation capabilities is gaining relevance in the field of sustainable construction and smart cities. When combined with HBIM and data from Internet of Things (IoT) devices, digital twins can provide valuable insights for optimizing energy systems, preserving building integrity, and enhancing visitor comfort in heritage buildings [15, 16].

This paper explores the use of an integrated digital workflow combining HBIM, indoor microclimate simulation, and LCA in the context of a heritage building in Kaunas, Lithuania: the Miko and Kipras Petrauskas House, designed in 1924 and listed in the National Cultural Heritage Register since 2001. The aim is to demonstrate how such a workflow can support data-driven, sustainable renovation strategies for heritage buildings. By applying this approach to a real case study, the research evaluates how digital tools can enhance indoor environmental quality and reduce embodied carbon while preserving architectural value and contributing to the broader development of digital twin frameworks in smart cities.

This study employed a multi-method, integrated workflow combining HBIM, indoor environmental monitoring, and LCA. The HBIM model not only documented the building's geometry but also served as a foundation for environmental assessments and material evaluations. By linking digital modelling, real-world environmental data, and life cycle impacts, the study aimed to inform renovation strategies that are both historically sensitive and environmentally optimized.

The following research questions guide this study:

1. How can HBIM be used to accurately document and digitally reconstruct a cultural heritage building to support sustainability-focused renovation planning?
2. To what extent can LCA, integrated with HBIM data, identify opportunities to reduce embodied carbon in heritage building renovation through material substitutions?
3. How can the combined use of HBIM, IAQ analysis, and LCA contribute to the development of digital twin frameworks for heritage buildings in the context of smart and sustainable cities?

2 Methods

This study employed a multi-method approach to assess the sustainability and indoor environmental quality of a heritage building using digital modelling tools. The methodology integrates digital capture (photogrammetry and laser scanning), HBIM, IEQ monitoring, and LCA into a connected workflow. This integrated approach ensures that digital documentation directly informs the evaluation of environmental conditions and material impacts, supporting data-driven renovation decisions.

2.1 Research Object

The object of the study is the Mikas and Kipras Petrauskas House, a cultural heritage building located in Kaunas, Lithuania. Constructed in 1924 and renovated in 2008, the building has been listed in the Lithuanian Cultural Heritage Register since 2001. It is characterized by interwar architecture and features reinforced concrete, silicate brick masonry, and wooden and metal structural components.

The first data logger was installed in office 313, which is occupied by three employees. The room is located on the third floor and has three south-facing windows. Employees reported discomfort during the cold season, often resorting to additional electric heaters. The device was positioned on a shelf at a height of 1.5 m, unobstructed by furniture or objects, and shielded from direct sunlight.

The second measuring device was placed in office 305, also located on the third floor. The room contains two workstations and serves as a passageway to a single-occupancy office. It features two skylights, with windows facing east. Employees have reported discomfort due to low air temperatures during the cold season and often use an electric radiator to warm the space. The measuring device was installed on a shelf along the interior wall at a height of 1.5 m, positioned to avoid obstruction by furniture or other objects and to prevent direct sunlight exposure.

The third device was installed in the children's education room located on the second floor. This room is used for educational activities involving children and remains unoccupied at other times. During sessions, the room accommodates up to 15 children. It is equipped with two windows and a door leading to a balcony, all facing northwest. The measuring device was positioned on a shelf along the external wall at a height of 0.5 m, free from obstructions and protected from direct sunlight.

The fourth measuring device was placed in the assembly hall on the building's ground floor. This space is used for weddings and other events, with the number of occupants varying depending on the occasion. At other times, the hall remains unoccupied. It features five windows, four of which face west and one facing southwest.

2.2 Data Acquisition and HBIM Model Development

The building was digitally captured using two complementary techniques:

- Photogrammetry was applied to define the external contours using aerial images captured via DJI Mavic 2 Pro.
- Terrestrial laser scanning (TLS) was used to generate the internal spatial configuration, employing a GeoSLAM ZEB Go LiDAR scanner.

The collected point cloud data were processed to generate a three-dimensional digital representation. Using Autodesk Revit, a HBIM was constructed by combining photogrammetric and LiDAR data with available renovation documents.

The model's accuracy was validated by comparing key geometric parameters—volume, height, width, and length between the HBIM model and the processed point cloud data obtained from photogrammetry and terrestrial laser scanning. Deviations were calculated as the absolute difference between HBIM-derived values and reference measurements, divided by the reference measurement and expressed as a percentage.

2.3 Indoor Microclimate Monitoring

To assess indoor environmental quality, in-situ measurements of air temperature, relative humidity, and CO_2 concentration were conducted in selected rooms during both warm and cold seasons. Measurements during the cold season were conducted from February 28, 2024, to April 17, 2024, and during the warm season from July 2, 2024, to August 24, 2024.

Measurements were taken using COMET U3430 sensors. The placement of the measuring devices remained consistent across both measurement periods. The selected instrument is capable of logging data at intervals ranging from one second to one day, with a storage capacity of up to 550,000 data points. The device's measurement accuracy is ±0.4 °C for temperature, ±1.8% for relative humidity, and ±50 ppm for CO_2 concentration. Prior to deployment, the instruments were fully charged, calibrated, and had their memory cleared in the laboratory. The data logging interval was set to 10 min.

2.4 Life Cycle Assessment

The Life Cycle Assessment was carried out using the One Click LCA software, which is specifically tailored for the building and construction sector. LCA methodologies assess environmental impacts across various stages of a product's life cycle, including "gate-to-gate" (processes within a manufacturing facility), "cradle-to-gate" (from raw material extraction to the factory gate), and "cradle-to-grave" (from raw material extraction through to end-of-life disposal).

In this study, the analysis focused on life cycle stages A1 to A3 for the analised building, where stage A1 encompasses the extraction of raw materials, stage A2 covers their transportation, and stage A3 assesses environmental impacts from the manufacturing process.

One Click LCA software integrates seamlessly with Revit-based building information models, allowing for the automatic extraction of material data and facilitating comprehensive environmental impact assessments. This integration supports informed and sustainable design decision-making.

3 Results and Discussion

This section presents the outcomes of the study, structured around the three main research questions: (1) the applicability and accuracy of HBIM for heritage documentation, (2) the potential of LCA-informed material substitution for reducing embodied carbon, and (3) the integration of digital tools for advancing digital twin development in the context of sustainable renovation.

3.1 HBIM Accuracy and Digital Reconstruction

The HBIM model of the Mikas and Kipras Petrauskas House was developed using a combination of photogrammetric and LiDAR data to accurately capture the building's geometry. Laser scanning was employed to document the interior spaces, while drone-based photogrammetry was used for the exterior (see Figs. 1 and 2). This integrated data collection approach eliminated the need for manual measurements, enabling efficient and precise transfer of the building's geometry into a 3D environment.

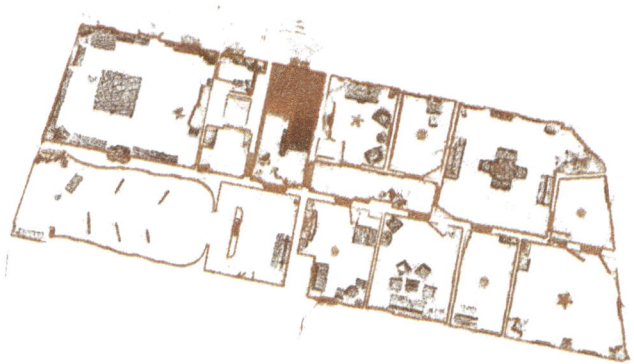

Fig. 1. The visible floor plan outline of the analyzed building from laser scanning [authors' image].

The drone captured numerous overlapping images of the building from various altitudes and angles to ensure comprehensive coverage. These images were processed using photogrammetry software to generate a detailed 3D model of the exterior. This model

Fig. 2. Photogrammetric model of the analyzed building [authors' image].

served as the basis for constructing a digital representation in Revit, including essential architectural components such as the façade, roof, windows, and other structural elements.

A comparison between the HBIM model and the original photogrammetric data confirmed its high geometric accuracy, with deviations in volume, height, width, and length remaining below 2%. The high-accuracy HBIM model provided a reliable digital platform not only for geometric documentation but also for structuring environmental monitoring activities and material impact analyses.

Beyond providing accurate geometric documentation, the HBIM model developed through photogrammetry and LiDAR scanning was a foundational platform for subsequent environmental assessments. The spatial detail captured enabled precise identification of indoor zones requiring environmental monitoring, optimizing sensor placement for IAQ measurements. Furthermore, accurate material quantification from the HBIM model enhanced the reliability of LCA calculations, particularly for embodied carbon estimations. In the context of heritage buildings, such integrated digital modelling is critical to ensure that environmental evaluations are based on comprehensive and non-invasive data acquisition, supporting informed, low-impact renovation strategies.

3.2 Indoor Microclimate

Indoor environmental quality was assessed through on-site measurements conducted in both cold (Table 1) and warm (Table 2) seasons across representative rooms of the Mikas and Kipras Petrauskas House. Parameters monitored included air temperature, relative humidity (RH), and CO_2 concentration, which serve as critical indicators of thermal comfort, air quality, and potential risk to heritage materials.

Thermal Conditions
During the cold season, indoor air temperatures were relatively stable and generally remained within acceptable comfort thresholds. In the warm season, temperatures increased moderately, which are acceptable for naturally ventilated historic buildings. While the results do not suggest thermal discomfort under standard occupancy, they do

indicate limited active thermal control raising concerns about performance consistency under more extreme weather or occupancy conditions.

Table 1. Summary data of measured values during the cold period.

Location	Average T, °C	Standard deviation T, °C	Average RH, °C	Standard deviation T, °C	Average CO_2, ppm	Standard deviation CO_2, ppm
313 office	19.8	1.88	36	5	634	247
305 office	23.3	1.11	26	5	561	174
Education room	19.4	1.28	33	7	455	103
Concert hall	20.9	1.07	31	5	552	221

Table 2. Summary data of measured values during the warm period.

Location	Average T, °C	Standard deviation T, °C	Average RH, °C	Standard deviation T, °C	Average CO_2, ppm	Standard deviation CO_2, ppm
313 office	25.3	1.6	50	5.3	447	66
305 office	26.6	1.6	44	5.5	457	55
Education room	25.3	1.6	48	5.1	428	46
Concert hall	23.8	1.5	54	7.8	404	44

Relative Humidity

The most critical finding applies to low relative humidity levels during the cold season. These levels are below the recommended threshold (typically 40–60%) for both occupant health and the preservation of historic building materials such as wood, plaster, and painted finishes. Prolonged exposure to such dry conditions may accelerate deterioration, especially in unsealed joints and porous surfaces.

In contrast, warm-season RH values ranged within acceptable limits. The seasonal fluctuation highlights the sensitivity of heritage interiors to outdoor climate and suggests insufficient humidity buffering by the building envelope or passive systems.

These results underscore a lack of active humidity control, particularly in winter, and reinforce the need to address RH in renovation strategies not only for comfort but also for preventive conservation of material integrity. While mechanical interventions may be limited in heritage contexts, these findings argue for introducing passive or minimally invasive humidification solutions where appropriate.

CO_2 Concentration

CO_2 levels remained below critical thresholds throughout the monitoring period, with average values under 1000 ppm, suggesting adequate air exchange under the current use conditions. However, several short-term peaks were observed during group occupancy, indicating potential risks under increased usage scenarios. These findings reflect adequate baseline air quality but highlight the need for ventilation evaluation if the building's functional program or visitor density is expanded in the future.

The insights gained from indoor climate monitoring highlighted the critical role of environmental performance considerations in selecting renovation materials, further supporting the integration of HBIM data into life cycle assessment workflows.

3.3   Life Cycle Assessment

During this study, an information model of the renovated cultural heritage building the Kipras and Mikas Petrauskas House, was developed. The model was created based on the most up-to-date information available regarding the layout of the building's spaces, the newly designed or updated structures during renovation, and the materials used, including their thicknesses.

The environmental impact analysis in this study covers stages A1, A2, and A3. Due to a lack of data, the subsequent stages (A4 to C4) were not assessed. As a result, the analysis focuses solely on the emissions related to the materials used, without considering emissions from construction, operation, or other lifecycle phases.

To evaluate the environmental impact of renovation materials, a life cycle assessment was conducted using One Click LCA software. Initial results showed that materials such as concrete (C30/37), aluminum partitions, and parquet flooring were among the highest contributors to embodied carbon (Fig. 3), with total CO_2e emissions reaching 108.6 tones.

Three targeted material substitutions were tested:

- Concrete C30/37 was replaced with a variant containing 10% recycled binders, reducing emissions from 30.6 to 20.9 tones (31.7%),
- Cork-based parquet flooring was substituted with MDF-based parquet, reducing emissions from 13 to 2.1 tones (83.9%),
- EPS insulation was replaced with an environmentally certified equivalent, decreasing emissions from 1.8 to 0.75 tones.

These changes reduced the total CO_2e emissions to 84.6 tones a 22.1% improvement. This illustrates the substantial potential of LCA-integrated design in selecting renovation materials that meet both environmental and conservation criteria.

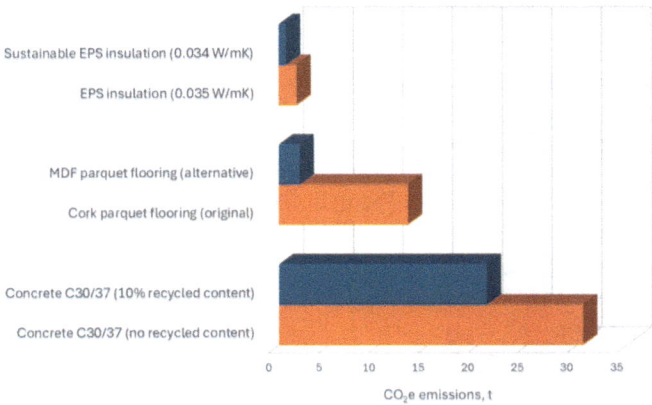

Fig. 3. Emissions from primary materials and they alternatives.

3.4 Towards Digital Twins for Heritage Sustainability

The integrated use of HBIM, IAQ analysis, and LCA in this study illustrates a multi-scalar digital workflow that not only documents and evaluates the existing conditions of heritage buildings but also lays the groundwork for future digital twin (DT) applications. In heritage contexts, precise digital modelling enables the non-invasive assessment of building conditions and facilitates environmental interventions that respect cultural value. Photogrammetry- and LiDAR-derived HBIM models support targeted IAQ monitoring by informing strategic sensor placement while also providing detailed material inventories necessary for accurate environmental impact assessments through LCA. Such integration enhances the potential for heritage buildings to achieve sustainability goals without compromising their historical integrity.

While the HBIM model developed in this study is static, it establishes a comprehensive digital foundation that can evolve into a dynamic digital twin by integrating periodically updated environmental and material health data. Future stages of digital twin development for heritage buildings would involve real-time indoor environmental monitoring, including parameters such as temperature, humidity, and CO_2 concentrations, as well as condition assessments of critical materials. These updates would enable predictive simulations, informed conservation strategies, and adaptive maintenance plans, aligning with digital twins' fundamental principle as dynamic, continuously evolving representations of physical assets.

This methodology offers a pathway toward more intelligent heritage building management, enabling stakeholders to evaluate scenarios, monitor environmental conditions, and plan low-carbon interventions without compromising cultural value.

4 Conclusions

This study demonstrates the practical application of an integrated digital workflow combining Heritage Building Information Modeling, indoor environmental monitoring, and life cycle assessment to support sustainability-oriented renovation of a cultural heritage building. The research aimed to evaluate how these tools could inform data-driven renovation decisions that both preserve historical value and reduce environmental impact.

First, the study showed that HBIM developed using photogrammetry and laser scanning can provide a highly accurate and interoperable digital representation of a heritage building. With less than 2% deviation in geometric parameters, the HBIM model proved suitable for integration with analytical tools, addressing the first research question related to accurate and efficient documentation for renovation planning. Moreover, the use of digital capture methods significantly reduced the time required for surveying and modelling while improving reliability, an important benefit in heritage contexts where existing documentation is often outdated, incomplete, or inconsistent with the actual built form.

Second, in response to the second research question, the LCA identified key contributors to embodied carbon in renovation materials. Targeted material substitutions including recycled-content concrete, MDF flooring, and low-GWP insulation resulted in a 22% reduction in total CO_2e emissions. This demonstrates how coupling HBIM-based material modeling with environmental impact analysis can guide more sustainable material choices in heritage contexts.

Third, the study revealed that indoor environmental measurements during the cold season indicated insufficient humidity control, with RH values falling below conservation thresholds. Although no simulations were presented in this paper, the IAQ analysis underscores the value of monitoring-based assessments in guiding future environmental interventions. When seen in conjunction with HBIM and LCA workflows, this supports the third research question: that integrated digital methods can lay the groundwork for data-informed, conservation-sensitive digital twins.

Overall, this work contributes to the evolving field of digital heritage and sustainable renovation by offering a replicable, integrated digital methodology. The workflow is particularly suitable for heritage buildings with limited or outdated documentation. However, limitations of the study should be acknowledged: (1) the HBIM model was static and not yet connected to real-time environmental data, (2) indoor climate monitoring covered two seasons but not a full annual cycle, and (3) material assessment focused on embodied carbon without directly analyzing material impacts on indoor air quality. Future work could expand this approach by developing operational digital twins, incorporating dynamic monitoring, and assessing broader sustainability indicators. Practitioners are encouraged to adopt phased strategies that prioritize early digital capture and ensure that sustainability considerations are embedded throughout the renovation process.

References

1. Srašinskaitė, E., et al.: Knowledge transfer in sustainable management of heritage buildings: Case of Lithuania and Cyprus. Academia.edu (2018) . https://www.academia.edu/113746 764/Knowledge_transfer_in_sustainable_management_of_heritage_buildings_Case_of_L ithuania_and_Cyprus
2. Gražulevičiūtė-Vileniškė, I., Šeduikytė, L., Liutikaitė, O.: Re-functioning of inter-war modernism buildings in Kaunas (Lithuania): the aspect of users' comfort. J. Sustain. Arch. Civ. Eng. **20**(1) (2017). https://sace.ktu.lt/index.php/DAS/article/view/19323
3. Kultūros Vertybių Registras: Nekilnojamųjų Kultūros Vertybių Paieška. https://kvr.kpd.lt/#/static-heritage-search. last accessed 17 Mar 2024
4. Chowdhury, M., et al.: Comprehensive analysis of BIM adoption: from narrow focus to holistic understanding. Autom. Constr. (2024). https://www.sciencedirect.com/science/article/pii/S0926580524000372
5. Lopez, A.J., Lerones, P.M., Llamas, J., Gomez-Garcia-Bermejo, J., Zalama, E.: A review of heritage building information modeling (H-BIM). Multimod. Technol. Interaction **2**(2), 21 (2018)
6. Olanrewaju, O., Babarinde, S.A., Salihu, C.: Current state of building information modelling in the Nigerian construction industry. J. Sustain. Arch. Civ. Eng. **28**(1) (2020). https://sace.ktu.lt/index.php/DAS/article/view/25142
7. Pupeikis, D., Morkūnaitė, L., Daukšys, M., Navickas, A.A., Abromas, S.: Possibilities of using building information model data in reinforcement processing plant. J. Sustain. Arch. Civ. Eng. **31**(2) (2021). https://sace.ktu.lt/index.php/DAS/article/view/27593
8. Pavloskis, M. (2021) Paveldo statinių tvarios konversijos modeliavimas. Master's Thesis, VGTU. https://vb.vgtu.lt/object/elaba:95620295/
9. Yin, H., et al.: Semantic localization on BIM-generated maps using a 3D LiDAR sensor. Autom. Constr. (2023). https://www.sciencedirect.com/science/article/pii/S0926580522005118
10. Cheng, C.P., et al.: Thermal performance improvement for residential heritage building preservation based on digital twins. Heliyon **10**(4) (2024). https://www.sciencedirect.com/science/article/pii/S235271022302466X
11. Prozuments, A., Borodinecs, A., Zemitis, J.: Survey based evaluation of indoor environment in an administrative military received facility. J. Sustain. Arch. Civ. Eng. **27**(1) (2020). https://sace.ktu.lt/index.php/DAS/article/view/26079
12. Ali, M., et al.: Heritage building preservation through building information modeling: reviving cultural values through level of development exploration. Semantic Scholar (2018). https://www.semanticscholar.org/paper/HERITAGE-BUILDING-PRESERVATION-THR OUGH-BUILDING-OF-Ali-Ismail/ca007a48dc2825b624e6f012044ba33d8b820031
13. Sanhudo, A., et al.: BIM framework for the specification of information requirements in energy-related projects. ResearchGate (2020). https://www.researchgate.net/publication/348 082321_BIM_framework_for_the_specification_of_information_requirements_in_energy-related_projects
14. Morsink-Georgali, F.Z., Kylili, A., Fokaides, P.A.: Life cycle assessment of flat plate solar thermal collectors. J. Sustain. Arch. Civ. Eng. **20**(1) (2017). https://sace.ktu.lt/index.php/DAS/article/view/18299
15. Cheng, H., et al.: BIM applied in historical building documentation and refurbishing. ResearchGate (2015). https://www.researchgate.net/publication/282524229_BIM_app lied_in_historical_building_documentation_and_refurbishing
16. Deng, M.: From BIM to digital twins: a systematic review of the evolution of intelligent building representations in the AEC-FM industry. J. Inf. Technol. Constr. (ITcon) (2021) . https://itcon.org/papers/2021_05-ITcon-Deng.pdf

Building Digital Twin – Digitalisation of the Thermo-accumulator Used for the Building's Heating

Tadas Zdankus[1]([✉]), Rao Martand Singh[2], Lazaros Aresti[3], Juozas Vaiciunas[1], and Sandeep Bandarwadkar[1]

[1] Kaunas University of Technology, Kaunas, Lithuania
tadas.zdankus@ktu.lt
[2] Norwegian University of Science and Technology, Gjøvik, Norway
[3] Cyprus University of Technology, Limassol, Cyprus

Abstract. When creating a digital twin of a building, all engineering system devices within the building must be digitised and integrated into the numerical model. This research focused on describing the processes occurring in a heat storage system, which was designed to accumulate heat during the summer period and utilise it by the building's heating during the cold period. This is a long-term acting heat storage system with a soil-type filler. Several stages of the thermal energy storage system operation were distinguished: charging, discharging, and heat retention. During all these processes, heat exchange with the environment occurs. Typically, it is heat loss to the environment. Experimental research was performed in field conditions to analyse the mentioned processes. After each charge, the heat was dissipated in the soil volume. The output signals of the sensors were recorded and analysed. It was noticed that two temperature measurement sensors are necessary to estimate the type of work regime and charge or discharge intensity. More sensors are needed to determine the amount of stored energy more accurately.

Creating a validated numerical model and comparing measured temperatures with simulated values at the same points enables a highly accurate assessment of the stored energy in the accumulator, as well as the description and forecast of parameter changes. By integrating the numerical model of the accumulator into the building's digital twin and combining it with building engineering systems models, it is possible to enhance the efficiency of the building's engineering systems and reduce energy consumption.

Keywords: Thermo-accumulator · Heat Storage · Heat Transfer · Digital Twin · Soil

1 Introduction

Numerical research methods have made a breakthrough in the field of scientific research [1]. Complex systems of differential equations have become surmountable, and the possibility appeared to create mathematical models of a process, phenomenon, or device [2].

© The Author(s) 2026
A. Jurelionis et al. (Eds.): BDTIC 2025, LNCE 775, pp. 123–133, 2026.
https://doi.org/10.1007/978-3-032-09040-9_11

That has dramatically facilitated the conduct of experimental research. After modelling various possible scenarios, only the most optimal cases are experimentally tested, thus saving equipment and time [3]. It is also possible to test critical or dangerous cases without conducting experiments. Using statistical analysis to analyse experimental research data, predict results, and evaluate various scenarios is possible [4]. The focus is on more serious challenges, such as creating digital twins. It means making the most realistic image of an object in a virtual space [5]. For this purpose, the virtual twin must be coherent with the real object. The interface may be realised by using the sensors, whose signals determine the state of the object and the ongoing processes [6]. Since the object is in a surrounding environment, it is appropriate to measure and know at least the environment's main parameters [7]. Monitoring changes in these parameters enables the accurate determination of the impact on the object and the prediction of what will happen after a specific time interval [8].

What are the benefits of digital twins? It is the possibility of the more efficient use of the object, its more effective operation, the extension of its life, the ability to avoid undesirable situations, to experience as little loss as possible, and so on. In general, on the one hand, this is a purely economic benefit; on the other hand, it is a sustainable interaction of the object with the surrounding environment [9].

In the civil engineering sector, it is necessary to consider the object's functions when making its digital twin. It may be a dam, as an example of the object, the functions of which are to withstand loads acting from the side of the pond, to stop the motion of groundwater, to extinguish the kinetic energy of the flow being passed, etc. On the other hand, a bridge must withstand dynamic loads, possible vibrations, etc. [10].

The digital twin of a building is understood as a complex image in virtual space, covering not only the structural integrity but also the heating, ventilation, and automatic regulation of air parameters, indoor lighting, plumbing, and other engineering systems. Interaction with the environment is also essential. The atmospheric air temperature, daylight hours, solar radiation intensity, and other factors must be known. If the definition of the building's environmental parameters, depending on seasonality, can be based on the stored data of meteorological stations (Construction Climatology), the inside parameters of the buildings must be measured in real time. It is not enough to observe the collected data; numerical simulation needs to be performed, and a way must be sought to optimise the operational properties of the building [7].

Collecting data makes it possible to assess not only the possible heat losses to the environment but also the efficiency of all engineering systems in the building. After numerical simulations have determined the optimal regime of the engineering system, it is possible to compare that with the real one and change the parameters of the controllers, leading the actual system to an optimal operational regime. By analysing the parameters and operation of the systems during the day, week, month, season, and year, it is possible to find a way to optimise the long-term operation of any engineering system and reduce energy consumption [11].

Managing the building's heating and cooling systems ensures the highest energy efficiency, thus reducing operating costs. Digital twins allow the simulation and optimisation of heat or coolness flows, heat accumulation, and dissipation while selecting the best methods and scenarios for the ventilation of the accommodations [11].

The investigation discussed in this article concerns an individual building with a living area of $100 \div 200$ m^2. Solar collectors produce heat for the building's heating. A water tank-type heat accumulator can be installed inside the building to distribute better heat for heating the building for one or a few days. The phase change materials should be used as the fillers for the thermal accumulator to extend the heat storage period to a week. A soil-type heat accumulator was installed below the building to extend the time to the interseasonal period.

This research aimed to digitise the processes in the heat accumulator and create a digital twin of a soil-type thermo-accumulator. The questions to be answered are: What needs to be known, and where and how are the parameters required to be measured to estimate the amount of stored energy? On the other hand, the thermo-accumulator must be integrated into the building's engineering systems and the digital twin of the entire building.

By supplementing the building digital twin with a digitalised thermo-accumulator operating on real-time data, modelling, optimising, and predicting the operation of the heating and cooling systems, it is possible to increase the efficiency of building heating systems and contribute to the broader goal of sustainable energy use. In a later step, going to the larger scale's heat storage could significantly contribute to the creation of smart cities.

2 Digitalisation of the Thermal Energy Storage System

The thermal energy of a constant-volume body is directly related to its internal energy, which in turn depends on its temperature. The temperature is often measured relative to a reference temperature, thereby defining the amount of energy in relation to the reference. In the case of a thermal storage system, the reference could be a value set by the energy consumer based on what the heat will be used for. In the case of building heating, the indoor air temperature that needs to be maintained is determined. It can be a fixed value of this temperature taken as a reference. On the other hand, it would be the value of the task signal for the temperature maintenance system controller, e.g., $T_{set} = 20\,°C$. Then, if the average temperature of the heat accumulator's ground filler T_{av_ac} is higher than the aforementioned task temperature T_{set} ($T_{av\_ac} > T_{set}$), the heat storage is charged with heat. In the case of the heat pump use, the temperature of the freezing point of water can be set as a reference ($T_{set} = 0\,°C$). Using the soil temperature surrounding the heat storage system below the ground surface ($h > 1$ m) is complicated as a reference temperature. This temperature depends on the season, and heat losses from the storage system can also affect it.

The amount of energy charged into the heat storage system E_{st} can be found by Eq. (1):

$$E_{st} = \rho V c_p \left(T_{av\_ac} - T_{set} \right). \tag{1}$$

where:

ρ – Density of the soil, kg/m^3.

V – Accumulator's volume, m^3.

c_p – Specific heat of the soil, J/(kg $\cdot$ K).

t – Time, s.

The process of the charge or discharge can be described by Eq. (2) (Fig. 1):

$$E_{st} = \rho V c_p \Delta T_{av\_ac} = k\rho V c_p \left(T_{t(j+1)} - T_{t(j)}\right). \tag{2}$$

where:

k – Empirical coefficient, dimensionless.

T_t – Temperature measured at the point of the accumulator, °C.

j – Time moment, s.

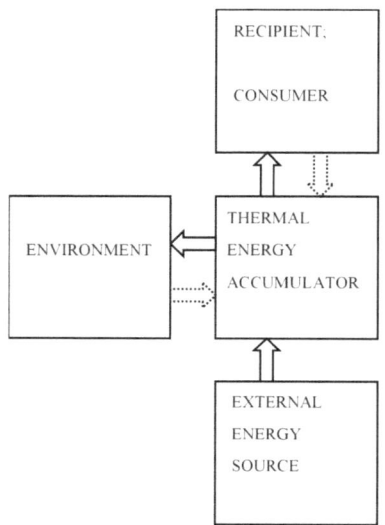

Fig. 1. Operating scheme of the heat storage system.

The change in the temperature of the soil filler over time can show what process is taking place: the temperature increase $\Delta T_{av\_ac}/t > 0$ means that the heat storage system is charging; $\Delta T_{av\_ac}/t \leq 0$ means discharge or heat loss E_{loss} to the environment. Heat loss to the environment occurs continuously. The soil surrounding the heat storage system acts as a secondary heat storage. It would be appropriate to measure its temperature T_{s_gr}. This would allow us to understand the state of the soil outside the heat storage system. A slow change in the temperature T_{s_gr} (when the depth $h \geq 1$ m) indicates seasonality, and a more intense change in T_{s_gr} indicates that the heat storage system has probably been charged or is charging. Then, the temperature of the soil filler increases, and the temperature difference $\Delta T_{loss} = T_{av\_ac} - T_{s\_gr}$ also increases. This means a possible increase in heat loss to the environment. When the heat storage system is discharged, the temperature difference ΔT_{loss} decreases. If the temperature of the accumulator's filler becomes lower than that of the ground outside, a change in the direction of heat spread is possible.

For the minimum requirement, the heat storage system's temperature should be measured at least at one point. However, a very pronounced inertia will be observed.

Therefore, measuring the temperatures of the system's filling at least at two points is recommended. One temperature sensor could be installed next to or at least closer in the distance of $l = 0.1 \div 0.2$ m to the energy source, and the other sensor could be installed in the cent e of the heat storage system's volume, limited by its walls and the heating circuit.

What rameters need to be known to describe the soil-type accumulator itself? First, the para ters of the soil, which was used as the filler, are needed: type, density, and moistu e fterwards, the heat capacity of the soil (the ability to store heat) and thermal condu iv y (a parameter required to calculate charge or discharge intensity, also related to heat lc s). Next are the heat storage system's capacity dimensions: height, length, and wi ltt Based on that, the volume may be found, and based on the volume, it is possibl: t compute the amount of energy that could be charged. The depth at which the heat accu nulator is located is needed as well. There should be no interaction with the groundwa er. Therefore, the level of the water table can be indicated. Depending on the depth, the type and moisture of the surrounding soil should also be defined. Structural features of the heat storage system: thickness, material, and properties of each layer of the thermo-accumulator's walls, if the walls were constructed.

In the case of high accuracy heat storage system's model preparation, the accumulator's thermal charging device must be described: the external energy source itself, the power of the energy source and its variation over time, the type of thermal charging device (electrical, heated fluid flow, etc.), the position of the charging device in the filler (bottom, inner part, center, top), the shape and dimensions of the charging device (tubes, tube loop, tube coil, flat surface - single-sided, double-sided, etc.). The energy extraction, i.e., thermal discharge device, must be described similarly.

3 Methodology

The experiments were conducted in field conditions. The heat storage system was placed underground at a depth of 0.2 m below the ground surface. The accumulator had the shape of a parallelepiped, whose internal volume was equal to $V = 0.25$ m^3 (Fig. 2). The horizontal cross-section was square, and the area (A) was equal to $A = 0.5 \times 0.5 = 0.25$ m^2. The height was equal to 1 m. The accumulator's volume was filled with sand of 10% moisture. Sand parameters were such: $\rho = 1845$ kg/m^3, $\lambda = 1.7 \div 2.0$ W/(m·K), $c_p = 1200 \div 1300$ J/(kg·K) [12].

Two heating devices were constructed; however, the first, located at the top of the heat storage system, was not utilised in this experimental research. Therefore, it is not described. The second heating device was made from heating tubes and covered by a flat steel surface. This device was placed at the bottom of the heat storage system. The electric resistance of the heating tubes was equal to $R = 71.4$ Ω. The electric current I was measured during the experiments by a special multimeter of the ESCORT 3136A type. Five temperature sensors, TJ1–Pt1000/A, were used for the temperature measurement and placed at a vertical distance equal to $h = 0, 0.1, 0.5, 0.9,$ and 1.0 m from the heated surface centre. Data Logger PT-104, connected to the computer, was used to read the sensors' signals.

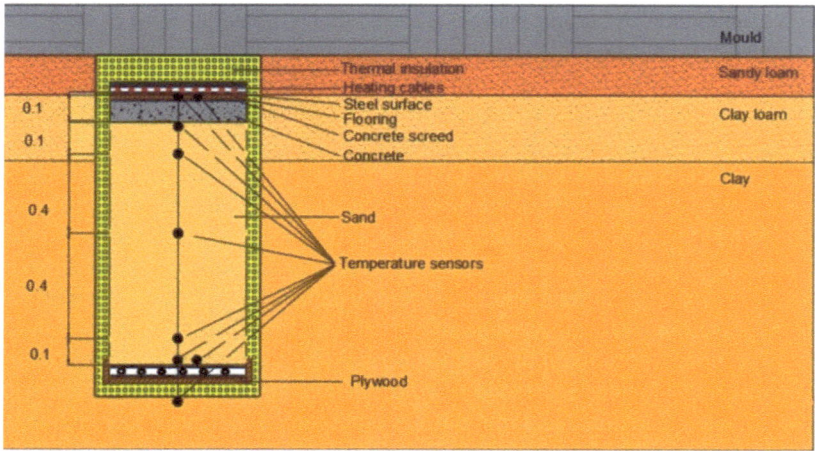

Fig. 2. Experimental setup.

The heat storage system's operation was simulated numerically using COMSOL software, version 6.1. A numerical simulation of a thermal accumulator constructed below the building (Fig. 3) was planned in the next step.

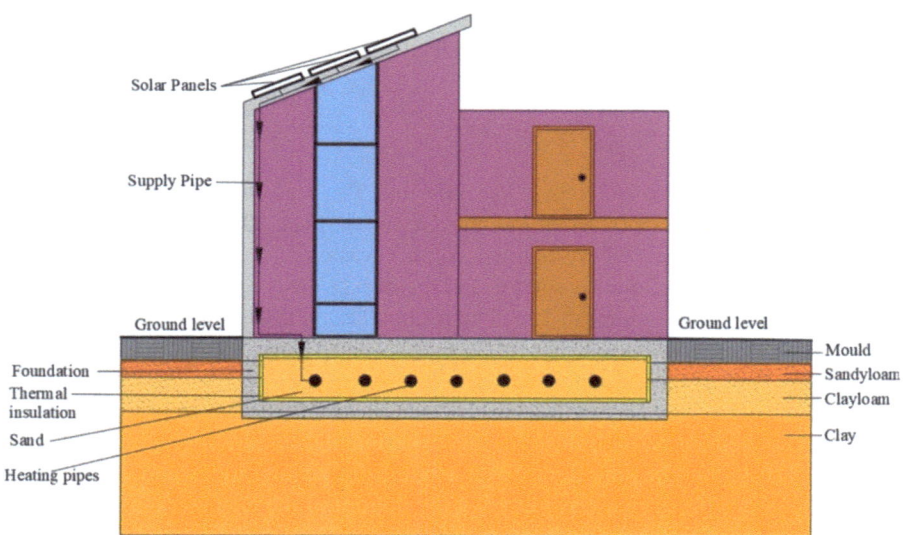

Fig. 3. Heat storage system constructed below the building.

4   Results

The results of experimental research on the heat storage system's charge are presented in Fig. 4. During this two-hour charge, the heating device consumed 1.39 kWh of electricity. When the heating device was turned on, the energy was used to heat the heating elements and the surface. Hence, the temperature sensor installed at the centre of the heated surface from the ground side began to respond after 72 s. The moment of connecting the heating device can be used as a reference point, but the ground filler begins to charge when the sensor responds. That can also serve as a reference point. When the heating device was disconnected, its temperature was higher than the ground. This meant that the thermal charge continued until the temperatures equalised. Therefore, temperature sensors are of priority.

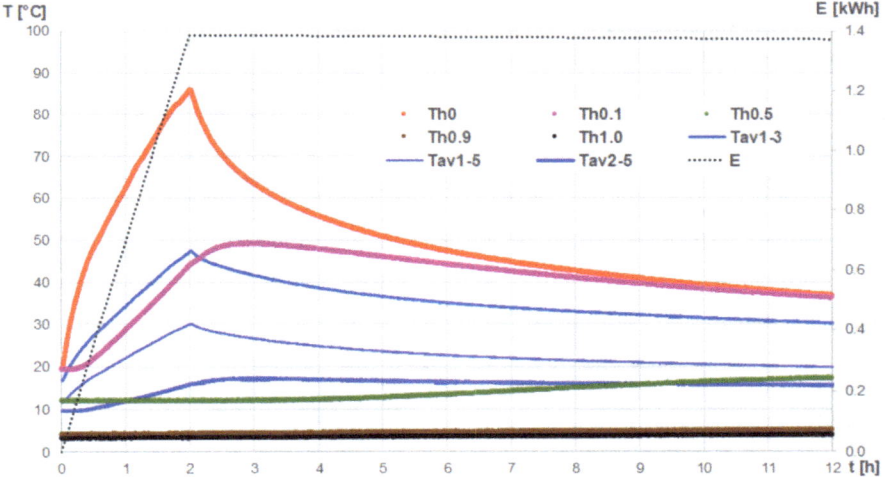

Fig. 4. Change in temperature at different points of the thermo-accumulator, an average temperature change depending on time, and the amount of energy charged to the accumulator.

As the heat spread to other soil layers, sensors at varying distances from the heating device measured the first temperature change at later times. During the experiment, the sensor installed at a distance of $h = 0.1$ m began to record the temperature increase after $t = 11$ min 24 s, and at a distance of $h = 0.5$ m, after two and a half hours.

Analysing the character of the temperature change of the heat storage system's soil filler, the temperature T_{h0} measured at the heating device quite accurately defines the time of thermal charging. At a distance of $h = 0.1$ m ($T_{h0.1}$), a delay in time was observed, and at a distance of $h = 0.5$ m, no change in temperature $T_{h0.5}$ was observed in the first two hours, corresponding to the accumulator's charging.

After the thermal charge was completed, temperature T_{h0} decreased. However, the heat from the more heated soil layer spread to cooler soil layers (Fig. 2). Almost an hour later, temperature $T_{h0.1}$ also began to decrease. However, temperature $T_{h0.5}$ increased because heat was dissipated throughout the filler and reached its centre.

Five temperature sensors were placed in the heat storage system's filler. Two average temperatures were calculated: $T_{av1-3} = (T_{h0} + T_{h0.1} + T_{h0.5})/3$ and $T_{av1-5} = (T_{h0} + T_{h0.1} + T_{h0.5} + T_{h0.9} + T_{h1.0})/5$. Significantly higher values of temperatures T_{h0} and $T_{h0.1}$ affected the nature of the variation of T_{av1-3} and T_{av1-5} and inaccurately described the processes occurring in the accumulator. According to the readings and T_{av1-3} and T_{av1-5} calculations, an erroneous conclusion may be drawn about the energy use in the accumulator, that is, discharge or heat loss to the environment.

By eliminating the readings of the first sensor and calculating the average filler temperature as follows: $T_{av2-5} = (T_{h0.1} + T_{h0.5} + T_{h0.9} + T_{h1.0})/4$, a more accurate description of the process may be obtained. However, the slight decrease in temperature T_{av2-5} after the end of the charge introduces uncertainties.

It can be stated that the first sensor, which measures the temperature T_{h0}, is needed to determine the start of the charge, heat exchange conditions, and charge duration. For a detailed determination of the processes occurring in the heat storage system, at least two sensors are required in the soil filler itself.

If the parameters of the thermal charge or discharge device were measured, it would be possible to compute the amount of energy put into the heat storage system. In our case, these were the electrical parameters of the heating device: resistance and electric current. In the case of heat carrier fluid usage, the flow rate and the inlet and outlet fluid temperatures should be measured. Combining this data with measurements of filler temperatures would allow for a much better assessment of the processes occurring in the accumulator. The heat loss should also be determined.

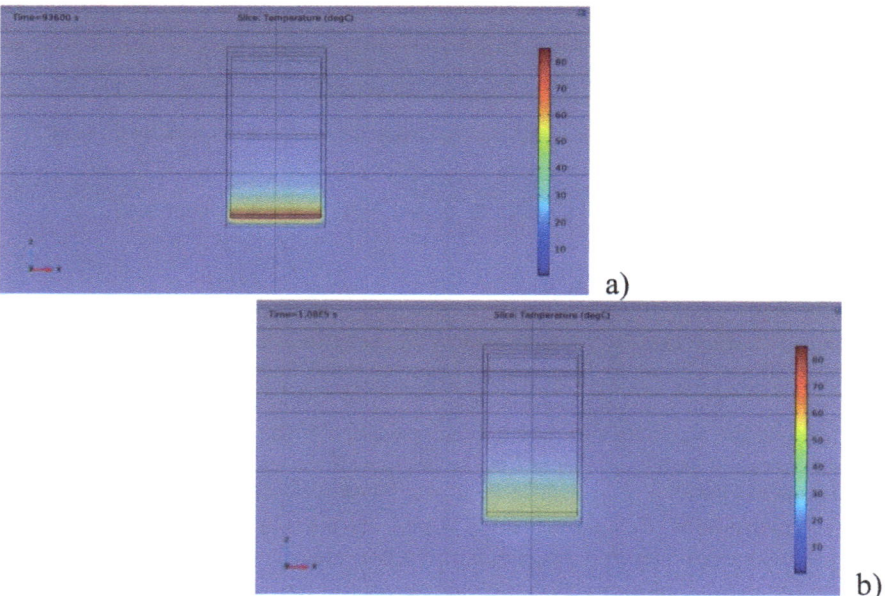

a)

b)

Fig. 5. The temperature profiles of the soil accumulator and surrounding ground in the case of the accumulator's model after $t = 2$ h (a) and 6 h (b).

It is possible to operate with the average soil-type heat accumulator temperature T_{av2-5}, using data from sensors located further from the charging and discharging devices. Then, evaluating the change in T_{av2-5} over time makes it possible to assess the processes in the filler. However, the assessment will be approximate without the first-mentioned sensor in contact with the heating or cooling device.

At least one, preferably several, temperature sensors, which can be placed on the outside of the heat storage system, above, below, and at its walls, are needed to assess heat losses to the environment.

Having a heat storage system model (Fig. 5) would make it possible to reduce the number of sensors, and the assessment of the processes would be much more accurate.

Temperature sensors placed in the filler of the accumulator should be connected to the data logger to enable the signal to be read and transmitted remotely. The temperature and the other sensors can be connected to a cloud-based database using communication protocols such as Modbus, MQTT, or BACnet. The collected data can be transmitted and processed using AWS IoT. The digital twin of the heat accumulator needs to be created and integrated into the digital twin of the building using BIM. We recommend using COMSOL Multiphysics, but ANSYS is also very helpful in simulating the heat storage system's thermal processes: the charge and discharge of the heat storage system and heat loss to the environment. After the model of the thermo-accumulator is created, it is possible to use AI, train machine learning models for specific tasks or write a program for the unique algorithms. The tasks may be different, such as optimising thermal storage based on weather forecasts and building occupancy. The predicted results can be checked with actual results after a certain time, ensuring the accuracy of the predictions. It is necessary to display heat accumulator performance in real-time. Mobile apps would be used to monitor and control the accumulator's operation remotely.

5 Conclusions

When developing a digital twin of a building that includes a thermal energy storage device, it is necessary to understand the principles of the heat storage system's operation. Mathematical models and numerical simulations enable us to determine the optimal number of temperature-measuring devices and place them most effectively.

Heat dissipation in the ground is a relatively slow process. Therefore, at least several temperature sensors are required to monitor, evaluate, and describe the processes in the soil-type heat storage system in real time. At least one sensor should be in contact with or near the thermal charging and discharging systems. An additional temperature sensor, which is 0.1–0.5 m away from the walls and devices placed in the ground, is required to assess the amount of energy stored in the ground. At least one temperature sensor is needed outside the heat accumulator to determine heat losses.

Creating a numerical model of the heat storage system enables the simulation of actual heat transfer within the accumulator. Additionally, measuring temperatures in real time improves the validation and calibration of the model. Then, it is possible to calculate the amount of energy accumulated in the heat storage system with great accuracy, predict the course of processes in the future and forecast the results. That allows the digitised heat storage system to be integrated into the digital twin of the building, which in turn can increase the efficiency of the building's engineering systems and save energy.

References

1. Iordache D.A., Sterian P.E.: Study of some complex systems by using numerical methods. Lecture Notes in Computer Science (including subseries Lecture Notes in Artificial Intelligence and Lecture Notes in Bioinformatics) 10961 (2018). LNCS: 539–559. https://doi.org/10.1007/978-3-319-95165-2_38

2. Foundations of Modelling and Simulation of Complex Systems. In: Proceedings of the Conference. 220054321 Foundations_of_Modelling_and_Simulation_of_Complex_Systems. https://www.researchgate.net/publication/. Last accessed 22 March 2025

3. Witkowski, K., Kudra, G., Wasilewski, G., Awrejcewicz, J.: Mathematical modelling, numerical and experimental analysis of one-degree-of-freedom oscillator with Duffing-type stiffness. Int. J. Non Linear Mech. **138**, 103859 (2022). https://doi.org/10.1016/J.IJNONLINMEC.2021.103859

4. The Beginner's Guide to Statistical Analysis|5 Steps&Examples. https://www.scribbr.com/category/statistics/?utm_source=chatgpt.com. (Accepted: 22 March 2025)

5. Fuller, A., Fan, Z., Day, C., Barlow, C.: Digital twin: enabling technologies, challenges and open research. IEEE Access **8**, 108952–108971 (2020). https://doi.org/10.1109/ACCESS.2020.2998358

6. IoT and Digital Twin Technology: Shaping the Future of Industry and Innovation - IoT Business News. https://iotbusinessnews.com/2023/11/14/08950-iot-and-digital-twin-technology-shaping-the-future-of-industry-and-innovation/. Last accessed 22 March 2025

7. Yan, J., Lu, Q., Li, N., Chen, L., Pitt, M.: Common data environment for digital twins from building to city levels. Autom. Constr. **174**, 106131 (2025). https://doi.org/10.1016/J.AUTCON.2025.106131

8. Van Dinter, R., Tekinerdogan, B., Catal, C.: Predictive maintenance using digital twins: a systematic literature review. Inf. Softw. Technol. **151**, 107008 (2022). https://doi.org/10.1016/J.INFSOF.2022.107008

9. Grieves, M., Vickers, J.: Digital twin: mitigating unpredictable, undesirable emergent behavior in complex systems. In: Transdisciplinary Perspectives on Complex Systems: New Findings and Approaches, pp. 85–113. Springer, Heidelberg (2017). https://doi.org/10.1007/978-3-319-38756-7_4

10. Liu, Z., et al.: Near-real-time carbon emission accounting technology toward carbon neutrality. Engineering **14**, 44–51 (2022). https://doi.org/10.1016/J.ENG.2021.12.019

11. Bortolini, R., Rodrigues, R., Alavi, H., Vecchia, L.F.D., Forcada, N.: Digital twins' applications for building energy efficiency: a review. Energies (Basel) **15**(19) (2022). https://doi.org/10.3390/EN15197002

12. Campanale, M., Moro, L., Siligardi, C.: Thermal properties of sands and their dependence on physical and environmental factors. Sci. Rep. **15**(1), 8352 (2025). https://doi.org/10.1038/S41598-025-93054-W

Overview of the Use of Co-creation Tools for the Design of Sustainable Buildings

Magdalena Okrzesik and Paris A. Fokaides$^{(\boxtimes)}$ ⓘ

Faculty of Civil Engineering and Architecture, Kaunas University of Technology, Kaunas, Lithuania
`paris.fokaides@ktu.lt`

Abstract. The transition toward sustainable building design increasingly emphasizes participatory approaches, where diverse stakeholders contribute to shaping solutions that balance environmental performance, social needs, and economic viability. This study provides an overview of the application of co-creation tools in the design process of sustainable buildings. Drawing on recent literature and practice-based examples, we examine how tools such as participatory design workshops, digital twins, virtual and augmented reality environments, and collaborative BIM platforms enable deeper engagement of users, designers, and decision-makers. The analysis identifies key benefits of co-creation, including improved user satisfaction, enhanced design adaptability, and stronger alignment with sustainability goals. It also explores challenges such as stakeholder coordination, data interoperability, and the need for capacity building in participatory methods. The study synthesizes findings across academic, professional, and policy domains to propose a typology of co-creation tools most relevant to sustainable architecture. The insights aim to inform both practitioners and researchers seeking to implement inclusive and effective design strategies in the built environment.

Keywords: Co-creation · sustainable building design · participatory design · Building Information Modelling (BIM) · Virtual Reality (VR) · user engagement

1 Introduction

The transition toward sustainable building practices is no longer driven solely by compliance with environmental standards or technological innovation. Increasingly, emphasis is being placed on participatory approaches that involve a wide range of stakeholders in the design and planning of buildings. Co-creation—defined as the collaborative process in which users, designers, engineers, and other actors jointly contribute to the development of solutions—has emerged as a key methodology for integrating social, environmental, and user-centric perspectives into sustainable architecture.

In the context of the built environment, co-creation tools include participatory workshops, stakeholder interviews, serious games, Building Information Modelling (BIM), digital twins, and immersive technologies such as virtual and augmented reality. These tools not only enhance user engagement and satisfaction but also improve the relevance

A. Jurelionis et al. (Eds.): BDTIC 2025, LNCE 775, pp. 134–141, 2026.
https://doi.org/10.1007/978-3-032-09040-9_12

and effectiveness of sustainability strategies by embedding user needs and local knowledge directly into the design process. They enable early-stage dialogue, continuous feedback loops, and shared ownership of outcomes, making them particularly suitable for projects that aim to be inclusive, adaptable, and environmentally responsive.

Despite the increasing attention to co-creation in recent years, there remains a gap in synthesizing how these tools are being systematically applied, evaluated, and integrated into the workflows of sustainable building design. To address this, the present study offers a structured review of relevant literature, based on the PRISMA (Preferred Reporting Items for Systematic Reviews and Meta-Analyses) methodology. An initial search yielded 72 scientific publications, from which 16 were selected through defined inclusion and exclusion criteria for full qualitative analysis. These studies form the core of the review and provide insight into the current state of co-creation practice within the sustainable construction domain.

The goal is to classify the co-creation tools used, identify their benefits and challenges, and provide a comprehensive overview to support future research and practice in collaborative and sustainable building design.

2   Methodology

This study applies the PRISMA (Preferred Reporting Items for Systematic Reviews and Meta-Analyses) methodology to conduct a systematic literature review focused on the use of co-creation tools in the design of sustainable and smart buildings. The aim was to identify and analyze peer-reviewed research that investigates co-creation methodologies within architectural or building-related design processes that explicitly target sustainability or user integration. For the bibliographic analysis, the Biblioshiny interface of the Bibliometrix R package was used. The figures presented were generated through Biblioshiny to visualize thematic trends, keyword co-occurrence, and country-topic relationships within the selected literature.

The dataset was retrieved from the **Scopus database**, widely recognized for its broad multidisciplinary coverage and reliable metadata. The search was conducted in **March 2025**, targeting studies published between **January 2012 and March 2025**. The keywords used were **"smart buildings," "co-creation,"** and **"design,"** combined using Boolean operators (*AND/OR*) and applied to titles, abstracts, and author keywords.

In the **Identification phase**, a total of **72 records** were retrieved. No duplicates were found in the dataset. During the **Screening phase**, titles and abstracts were assessed for thematic relevance, resulting in the exclusion of 29 articles that were unrelated to the built environment, co-creation methods, or sustainability in building design. This yielded **43 studies** for full-text review.

The **Eligibility phase** involved applying inclusion criteria:

- The study must involve buildings or built environment projects;
- It must explicitly incorporate co-creation or participatory design tools;
- The research must describe or evaluate the use of these tools;
- It must be situated within a sustainability or smart building context.

After applying these criteria, **16 studies** were included in the **qualitative synthesis**. These studies form the analytical basis of the review, where co-creation tools are categorized, their use cases documented, and their role in promoting sustainable building outcomes evaluated. The 16 studies selected through the PRISMA process are listed in the references section of this paper. The PRISMA flow table summarizes the selection process (Table 1).

Table 1. Summary of PRISMA Selection Process.

Stage	Number of Records
Records identified through database searching (Scopus)	72
Records after duplicates removed	72
Records screened (title and abstract)	43
Full-text articles assessed for eligibility	25
Studies included in qualitative synthesis	16

3   Results and Discussion

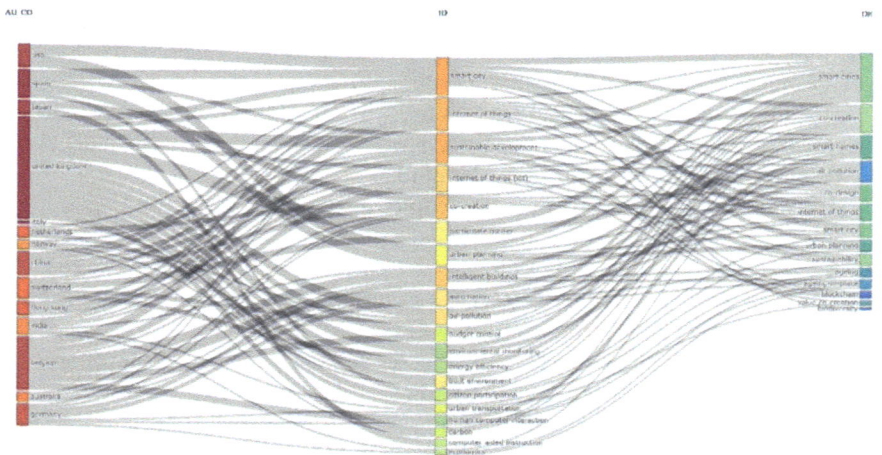

Fig. 1. Sankey diagram linking countries, keywords, and application domains in co-creation research.

Figure 1 presents a Sankey diagram mapping country contributions to keywords and their downstream application domains. It highlights the dominant role of countries like the **USA, UK, Japan, and Italy**, which lead the scholarly production around concepts such as *smart city*, *internet of things*, and *co-creation*. The visualization makes clear that **co-creation serves as a bridging term**, connecting diverse research fields such as urban planning, smart homes, and sustainable development. This suggests that co-creation is not limited to stakeholder workshops or early design phases, but is increasingly adopted as a system-level methodology across the entire building lifecycle—from concept to implementation and even into use-phase governance.

What is particularly notable in Fig. 1 is the strong flow from *co-creation* to *co-design* and *smart homes*, reinforcing the hypothesis that participatory practices are increasingly embedded in residential and user-interactive domains. The presence of terms like *value co-creation*, *urban planning*, and *air pollution* as downstream applications indicates that co-creation is also being leveraged to integrate environmental and policy objectives into the built environment, aligning with goals of sustainable development.

Figure 2, the word cloud, highlights **dominant and frequently co-occurring terms** in the reviewed literature. *Sustainable development*, *smart city*, *internet of things*, *intelligent buildings*, and *co-creation* emerge as the most prominent. This confirms the increasing convergence of digital and participatory paradigms in building design. Terms like *human computer interaction*, *environmental monitoring*, and *citizen participation* signal that researchers are moving beyond technical performance metrics to incorporate user-centered and societal perspectives. Importantly, **co-creation is surrounded by a set of complementary terms such as** *automation*, *budget control*, *urban planning*, **and** *energy efficiency*, suggesting that the practice is maturing into a structured and outcome-oriented design philosophy.

The presence of *computer-aided instruction* and *virtual reality*—though smaller in scale—points to an **emerging interest in immersive tools**, which support learning and visualization within participatory design processes. These tools, while not yet as central as the dominant terms, are gaining traction as facilitators of real-time user interaction, experiential feedback, and iterative refinement in co-design settings.

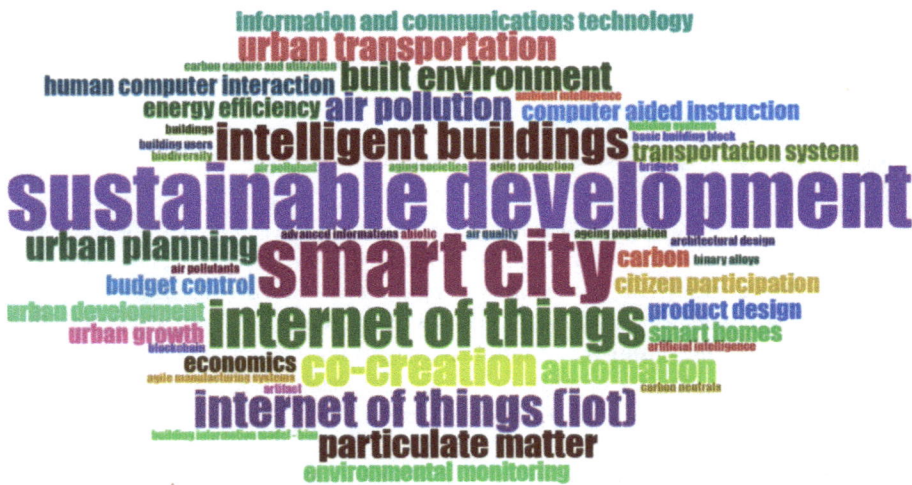

Fig. 2. Word cloud showing dominant terms in co-creation and sustainable building literature.

Figure 3, the thematic map, provides a two-dimensional representation of how central (relevant) and developed (dense) different research themes are within the field. The cluster in the upper-right quadrant, comprising **"co-creation," "built environment," and "carbon,"** is especially relevant. These are **well-developed and increasingly integrated themes**, representing domains that are not only rich in academic content but also vital to interdisciplinary research and practical application. Their location in the quadrant suggests that co-creation is transitioning from an emerging method to a well-established pillar within sustainable building practices.

By contrast, the lower-left quadrant includes *air pollution* and *urban development*, themes which may be either emerging or declining. This shift suggests that **the field is evolving away from siloed environmental metrics toward more systemic, participatory frameworks**—with co-creation and smart technologies at their core.

Meanwhile, the cluster labeled as "motor themes" in the top-right corner—*smart city, internet of things*, and *sustainable development*—reinforces that these are the anchors of current scholarly attention. The relative position of *co-creation* just outside this dominant cluster further emphasizes its growing recognition and strategic alignment with broader goals in digital urban transformation and sustainability transitions.

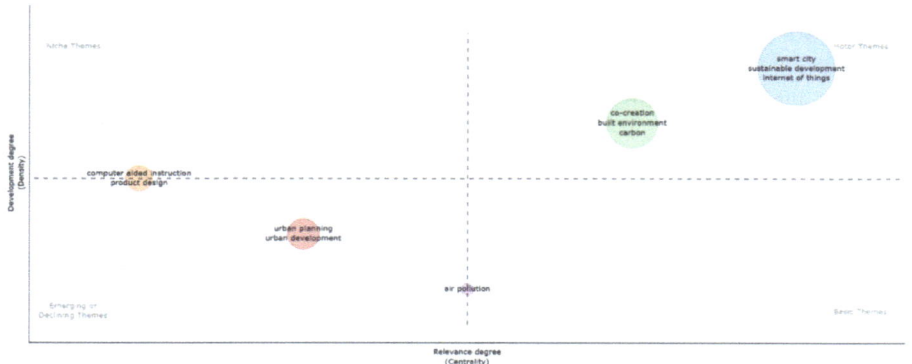

Fig. 3. Thematic map of research clusters by centrality and development within the field.

Figure 4, a co-word network focusing on immersive design and participatory technologies, adds nuance to this evolution. The centrality of *virtual reality* and *co-creation*, surrounded by terms such as *participatory design*, *user-centered design*, *design process*, *immersive environment*, and *e-learning*, provides compelling evidence that **Virtual Reality (VR) is rapidly becoming a methodological enabler of co-creation**. VR is situated at the core of networks that emphasize experiential design, collaborative interaction, and architectural simulation, supporting use cases ranging from civic engagement to energy-aware user behavior modeling.

Moreover, the visual clustering of VR with *training simulation, students, qualitative research*, and *real-world* applications underscores its dual role as both a research and educational tool. This positions VR as an accessible medium not only for professional design teams but also for diverse user groups, including future occupants, stakeholders, and even policymakers. By enabling stakeholders to visualize, explore, and modify spatial concepts in real time, **VR strengthens the feedback loop between users and designers**, turning abstract sustainability objectives into tangible design actions.

Taken together, the four figures support a clear narrative: **co-creation is a mature and central theme in sustainable building research**, with growing influence in the digital and smart design domains. While many traditional tools and participatory practices remain relevant, the bibliometric evidence strongly indicates that **Virtual Reality is emerging as a powerful and increasingly integrated method** within this ecosystem. Its capacity to simulate, engage, and co-create design solutions places it at the frontier of participatory architecture, especially in smart and sustainable buildings where complexity and stakeholder diversity demand new forms of communication and collaboration.

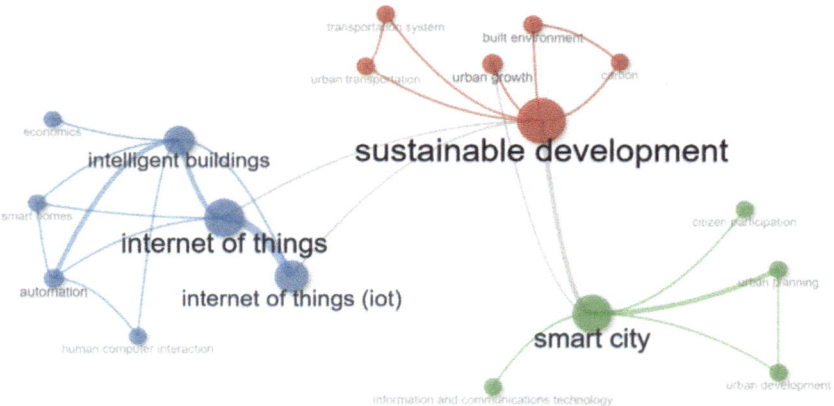

Fig. 4. Co-word network highlighting the role of virtual reality in participatory design.

4 Conclusions

This study provides a structured overview of co-creation tools used in the design of sustainable buildings, with insights drawn from a targeted bibliographic analysis of 16 studies selected through the PRISMA methodology. The findings confirm that co-creation is now widely recognized as an essential component of sustainable and smart building design, enabling the integration of stakeholder knowledge, behavioral insights, and context-specific needs into early planning and decision-making processes. Tools such as participatory workshops, stakeholder consultations, serious games, and collaborative digital platforms like BIM and digital twins are increasingly embedded in architectural workflows, enhancing transparency, adaptability, and design acceptance. The thematic clusters and keyword analyses indicate a well-developed and maturing research field that aligns co-creation with strategic goals in energy efficiency, climate resilience, and social inclusivity. As sustainability targets become more ambitious and the built environment more complex, co-creation methodologies will play an increasingly strategic role in ensuring user engagement and long-term performance outcomes. Among the variety of tools examined, Virtual Reality (VR) was identified as an emerging method with unique potential to enhance user experience and spatial understanding in co-design processes. Continued integration of co-creation into mainstream practice will be vital in shaping resilient, inclusive, and future-ready buildings.

References

1. Nidam et al. (2025). Practicing data inclusion: Co-creation of an urban data dashboard. *Environment and Planning B: Urban Analytics and City Science*, 52(1)
2. Zhang et al. (2024). Exploring the nexus of smart technologies and sustainable ecotourism: A systematic review. *Heliyon*, 10(11)
3. Sameer et al. (2024). Toward smart sustainable cities: assessment of stakeholders' readiness for digital participatory planning. *Archnet-IJAR: International Journal of Architectural Research*, 18(4)
4. Theodoropoulou et al. (2024). Glare-based control strategy for Venetian blinds in a mixed-use conference space with fully glazed facades. *Journal of Building Engineering*, 82
5. Berigüete et al. (2024). Digital Revolution: Emerging Technologies for Enhancing Citizen Engagement in Urban and Environmental Management. *Land*, 13(11)
6. Royo-Vela et al. (2024). The role of value co-creation in building trust and reputation in the digital banking era. *Cogent Business and Management*, 11(1)
7. Zwick et al. (2024). Examining the Smart City Generational Model: Conceptualizations, Implementations, and Infrastructure Canada's Smart City Challenge. *Urban Affairs Review*, 60(4)
8. Florentin et al. (2024). Facilitating citizen participation in greenfield smart city development: The case of a human-centered approach in Kashiwanoha international campus town. *Telematics and Informatics Reports*, 15
9. Violano et al. (2024). From Self-Reliant to Sufficiency Design: Predictive and Forecasting Features of Technology Approach. *Lecture Notes in Networks and Systems*, 1189 LNNS
10. Harra et al. (2024). Developing sustainable and ethically valid gerontechnology: A comprehensive approach to aging, diversity, and participation. *Gerontechnology*, 23
11. Huang et al. (2024). Application Models and Innovative Approaches of Smart Libraries from the Perspective of MR Technology. *Journal of Library and Information Science in Agriculture*, 36(9)
12. Ma et al. (2024). Developing a Manufacturing Industrial Brain in a Smart City: Analysis of fsQCA Based on Yiwu Knitting Industry Platform. *Buildings*, 14(5)
13. Zhang et al. (2023). Shaping a Smart Transportation System for Sustainable Value Co-Creation. *Information Systems Frontiers*, 25(1)
14. Badawi et al. (2023). Towards a Value Co-Creation Process in Collaborative Environments for TVET Education. *Sustainability (Switzerland)*, 15(3)
15. Slingerland et al. (2023). Beyond human sensors: More-than-human Citizen Sensing in biodiversity Urban Living Labs. *ACM International Conference Proceeding Series*
16. Kormann-Hainzl et al. (2023). Smart Regions in the Age of the Citizen Developer: A Service-Oriented Engineering Approach. *ACM International Conference Proceeding Series*

Knowledge-Based Configuration Expert System for Deep Renovation Planning of Buildings

Joosep Viik(✉) [iD], Ergo Pikas [iD], and Targo Kalamees [iD]

Department of Civil Engineering and Architecture, Tallinn University of Technology,
Ehitajate Tee 5, 19082 Tallinn, Estonia
joosep.viik@taltech.ee

Abstract. Renovation is a general solution for reducing building stock emissions and energy consumption. While technical competence and digital tools exist, the renovation process is hindered by various barriers, particularly for non-expert renovation initiators (e.g., apartment building association representatives, building managers). This study proposes a knowledge-based configuration expert system (KBCES) concept as a solution to support non-experts in early-stage deep renovation planning. Using the Design Science Research (DSR) methodology, a system architecture and prototype Renokratt were developed and evaluated for typical Estonian apartment buildings.

Renokratt was tested in a lab environment with results indicating that it can effectively support early-stage renovation planning, provided further development. The evaluation highlighted the importance of reliable building data. The study concludes that KBCES can bridge the gap between expert knowledge and end-user needs, improving the quality and pace of renovation planning while supporting broader climate goals.

Keywords: deep renovation · expert systems · knowledge-based systems · configuration systems · apartment buildings · building typology

1 Introduction

The European Green Deal has prioritized building renovation as a key strategy for reducing environmental impact and energy consumption, with Estonia aiming to renovate all pre-2000 buildings to at least energy performance certificate class "C" by 2050 [1]. The renovation of existing buildings, particularly Soviet-era apartment buildings, is a critical challenge in Estonia's efforts to reduce emissions.

Deep renovation is a complex process and involves multiple stakeholders, with non-expert renovation initiators often lacking the knowledge to effectively plan and articulate renovation requirements [2, 3]. This can lead to inconsistent quality of design tasks and challenges in the construction process. "Non-professional clients need comprehensive support that they can use independently and at their convenience" to make renovation planning more efficient [2].

Existing support systems are often designed for industry professionals. For example, Kamari et al. [4] presented a system to assist architects and engineers in designing

© The Author(s) 2026
A. Jurelionis et al. (Eds.): BDTIC 2025, LNCE 775, pp. 142–154, 2026.
https://doi.org/10.1007/978-3-032-09040-9_13

energy-efficient renovation scenarios, followed by research on a domain model and planning framework for performance-based design evaluation [5]. However, such tools may not adequately address the needs of non-expert renovation initiators such as apartment association representatives.

This study addresses these challenges by introducing a renovation planning tool for non-expert renovation initiators. Its primary objective is to support users configure renovation scenarios, assess the impact of scenarios, and articulate clear renovation requirements for design specialists. This study is needed to formalize renovation domain knowledge in a way that enables non-experts to independently navigate the early stages of deep renovation planning.

2 Research Methodology

This study implements the **Design Science Research (DSR)** methodology as outlined by Peffers et al. [6], which provides a structured process for developing and evaluating innovative artefacts to solve relevant practical problems with academic novelty. The structure of the study follows the DSR methodology phases:

1. **Problem Identification and Motivation:** An extended theoretical background is presented, and the problem is defined.
2. **Define Objectives and System Requirements**: The system objective and requirements are defined based on the FURPS [7] methodology.
3. **Design and Development**: The system architecture was developed and described. The proof-of-concept prototype, Renokratt, was developed.
4. **Demonstration**: The prototype was tested qualitatively with relevant renovation stakeholders in a lab environment. Testing focused on validating the functional requirements set for the system.
5. **Evaluation**: Assesses how the solution addresses identified problems, meets set requirements, and supports broader research objectives.
6. **Communication**: The results are documented and presented, including, for example, this paper.

3 Problem Identification and Motivation: Research Background

The Energy Performance of Buildings Directive (EPBD) positions building renovation as a key strategy for reducing emissions in the EU. There are approximately 27,000 apartment buildings in Estonia, with Soviet-era buildings, made of reinforced concrete panels, blocks, and bricks, comprising the largest portion [8]. The Estonian renovation strategy targets the renovation of ~14,000 apartment buildings, which requires a significant increase in renovation rates [1]. Key challenges that hinder the success of renovation projects in Estonia are outlined in Table 1.

Among these challenges, no. 1 to 4 are key in the early planning phase. Because the renovation initiator—a building board member, manager, owner or other interested party—is typically unfamiliar with construction, the Estonian renovation grants require the involvement of a technical renovation consultant (TC) [10], who manages the whole

Table 1. Renovation Challenges

#	Challenge	Description	Src.
1	Awareness	Clients don't know why, how to renovate or where to start.	[3]
2	Undervaluing project design	Clients focus solely on the construction, neglecting the importance of the project design phase. BIM design is underused, increasing the risk of design errors.	[2]
3	Non-expert client	Clients are unfamiliar with construction and lack the expertise to understand how design choices affect outcomes.	[2]
4	Design task quality	Clients struggle to define clear renovation requirements and articulate design tasks for designers.	[2]
5	Financial capacity	There is insufficient funding to fulfil regulatory requirements, and other critical work is postponed.	[3]
6	Lack of quality criteria	Procurement processes for both design and construction do not include quality criteria, resulting in lowest-cost tenders that compromise quality.	[2]
7	Omission of works	Key upgrades may be omitted depending on the renovation goal, e.g., EP improvements are omitted when the goal is to improve functionality or safety.	[3]
8	Non-holistic approach	Clients often prioritize just one of the three pillars—sustainability, health, or performance—and consider only short-term perspectives (20–30 years).	[3]
9	Unclear role of local governments	The role of municipalities in coordinating neighborhood-level renovation is unclear. Regional development is not considered enough.	[2, 3]
10	Cyclical public funding	Dependence on public funding cycles hampers business development, industrial and neighborhood-level renovations, reducing scalability and increasing per-unit costs.	[3]

(continued)

Table 1. (*continued*)

#	Challenge	Description	Src.
11	Regional imbalance of support	Apartment buildings outside urban centers need more financial and advisory support than those inside. The subsidy is often insufficient in low-value areas, where bank loans are harder to obtain.	[1]
12	Fragmented sector	Renovation experiences vary across municipalities and companies, leading to low overall productivity.	[2, 9]
13	Shortage of specialists	There is a general shortage of renovation professionals, particularly project design specialists.	[2, 9]
14	Superficial stakeholder engagement	The engagement of contractors in the development of renovation processes and technologies is superficial and irregular and lacks meaningful follow-up.	[2]
15	Fragmented feedback loops	Experience from clients and contractors is not shared across the sector. Feedback mechanisms between stakeholders are unstructured.	[2]
16	Regulatory barriers	Regulations do not reflect the greater complexity of deep renovation compared to new builds. Building permits often expire before construction is completed.	[2]

renovation process and handles tasks from design task creation to grant applications, procurements and contract negotiations [10].

But even with TC support, the key challenges persist. TCs themselves have highlighted the need to better understand industrial renovation and client options [2]. Providing training for non-expert clients is impractical and these challenges should be addressed in other ways, such as developing digital tools that incorporate previous renovation experiences [2]. It was concluded in [2] that "non-professional clients need comprehensive support that they can use independently and at their convenience". This support should help reduce both the actual and perceived complexity of renovation while better explaining the renovation process and addressing relevant pre- and misconceptions [2].

Digitalization offers significant potential for improving renovation efficiency, with digital and data-based solutions enabling simulation of various renovation scenarios and optimization of renovation measures [9]. A system that makes expert knowledge more accessible to non-experts has potential to significantly simplify renovation planning and help meet the growing renovation demand.

4   Define Objectives and System Requirements

The solution must address challenges no. 1 to 4 and support non-experts in configuring renovation scenarios, assessing the impact of scenarios and articulating clear renovation requirements for design specialists, while being usable independently and at the convenience of the renovation initiator [2]. The solution is a system that must create a set of requirements for the building that clearly and concisely describes the desired renovation outcome and can serve as a foundation for the subsequent renovation stages.

The system requirements are defined using the FURPS methodology [7] which categorizes requirements into: functionality, usability, reliability, performance, and supportability. The first category (F) describes functional requirements and the latter (URPS) non-functional requirements.

Based on the research objectives and the challenges of initiating renovations, the following **functional requirements** (FRs) were set. The system must:

1. Identify baseline knowledge about the building, including open data, the building's archetype, geometry, and statistical parameters.
2. Enable users to verify and specify baseline building information and enhance the dataset with as-is building knowledge.
3. Identify suitable renovation measures for the building.
4. Enable selection between different renovation measures and describe the advantages and disadvantages of different choices.
5. Determine whether the selected renovation measures meet technical, legal, financial and social requirements and principles set for the building.
6. Assess the impact of selected renovation measures on the building:
   a. Calculate estimated construction costs and changes in residents' running expenses,
   b. Calculate the energy performance (EP) of the renovated building.
   c. Calculate the greenhouse gas emissions (GHG) of the renovation.
7. Generate a comprehensive and sharable design task document describing the configuration based on the choices made.

While this study focuses on FRs, non-functional requirements (NFRs) are equally important for real-world applicability. As highlighted in software engineering literature, "real-world problems are more non-functionally oriented than they are functionally oriented" [11]. In other words, it is the NFRs that determine how effectively the system can address renovation challenges.

5   Design and Development: KBCES for Deep Renovation Planning

5.1   System Architecture

The **Knowledge-based Configuration Expert System** (KBCES) concept is suggested as a response to the challenges in initiating renovation projects. It is a synthesis of expert systems, knowledge-based systems and configuration systems.

Expert Systems (ES) are systems, where task-specific knowledge is transferred from a human to a computer, which then acts as a replacement to a human consultant, giving advice and explaining the logic behind the advice [12].

Knowledge-based Systems (KBS) have developed as "attempts to understand and initiate human knowledge in computer systems" [12]. KBS are described by the distinction of a knowledge base and an inference mechanism [13].

Configuration Systems (CS) are systems that assemble a tailored solution by selecting and combining predefined component types according to user requirements and domain-specific constraints. [14].

A KBCES is a synthesis of these systems and the architecture of it is based on the key components of a KBS: a **user interface**, a **knowledge base** and an **inference engine** [13]. These key components are described under the next headings and the architecture of the system is visualized in Fig. 2, where the classic expert system architecture is expanded with the renovation domain context.

5.2   User Interface

The user interface must guide non-professional users through the system, while providing sufficient information required to make informed renovation decisions. The configuration process is divided into six different stages:

1. **First Contact**: The user lands at the system and the system's purpose and capabilities are introduced.
2. **Input Building Address**: The user identifies the building to renovate.
3. **Data Validation**: The user can correct, validate and add to the building dataset. The user must provide additional information about previous renovations and the as-is building condition.
4. **Renovation Strategy Configuration**: Users can choose between and compare specific renovation measures tailored to the identified building.
5. **Configuration Analysis**: The system calculates specific indicators for the renovation configuration that support decision making.
6. **Export Results**: The system generates a sharable summary of the configuration which provides a basis for the subsequent renovation phases.

5.3   Knowledge Base

The knowledge base formalizes expert knowledge within the computer system [13]. It encompasses all relevant information pertaining to the building and the renovation process. It consolidates domain-specific knowledge and rules from various sources into one place, enabling new knowledge to be inferred from them.

The composition and structure of the knowledge base are context-dependent, varying across different national frameworks, regulatory environments, and local typologies. In the context of Estonian typical Soviet-era apartment building renovation, the knowledge includes data from the Estonian Address Data System (ADS), the Estonian Topographic Database (ETD) including the Estonian Digital Twin, Estonian Building Registry (EBR), building typology knowledge [8], renovation measures knowledge, renovation process knowledge and user input.

Open data sources (such as ADS, ETD, EBR) provide baseline information about buildings in Estonia. However, this data can be incomplete or even incorrect. This problem is especially acute in EBR, where the responsibility for filling data fields has been left to non-experts. The most reliable fields in EBR are those related to building ownership and real estate value (e.g., building's net area) [8].

The open data is complemented with **building typology** knowledge. For example, E. Iliste [15] developed a method to systematically determine the general archetype of Estonian Soviet-era stone apartment buildings. Based on this typology, specific statistical average parameters (e.g., proportion of windows on the façade, thermal properties of walls and roofs) can be assigned to the building.

To formalize the renovation measures, their relationships and constraints (exclusions and dependencies) in the knowledge base, a **configuration model** is used, which is a representation of said components and constraints. The configuration model enables to formalize renovation rules such as:

- R1: IF the building's exterior wall is insulated AND the building's windows are replaced, THEN the building's air leakage rate decreases.
- R2: IF the building's air leakage rate decreases, THEN a mechanical ventilation system must be installed in the building.
- R3: IF a ventilation system with heat recovery is installed in the building, THEN the building needs both supply and exhaust ventilation ducts.
- R4: IF the building has a flat roof, THEN a pitched roof structure type cannot be installed on the building.

For example, when implementing R3, ventilation unit subtypes with and without heat recovery can be created for parent type (Fig. 1).

In the configuration model, it is important to distinguish between component types (classes) and specific component type instances. A component type is, for example, a wall, which can be of the subtype exterior or interior wall. A specific instance is a component type whose attributes have assigned values and symbolizes a specific renovation measure.

Component type instances are stored in the **renovation measures database**. The renovation measures database is a collection of different specific choices (e.g., wall type instances, ventilation system instances) from which the user can choose from. The rules formalized in the configuration model are applied to the renovation measure component type instances, which filter user choice options.

The person responsible for acquiring system knowledge and developing and managing the knowledge base is called a **knowledge engineer**. The knowledge engineer and domain expert can be the same person, but it is important that the knowledge engineer is familiar with knowledge-based system technologies. [14].

5.4 Inference Engine

The inference engine is the part of the knowledge-based system that applies logical rules to existing building knowledge and infers new knowledge. It consists of three conceptually distinct parts: (1) interpretation module, (2) explanation module, and (3) solving module. The architecture of the KBCES for deep renovation has been visualized in Fig. 2.

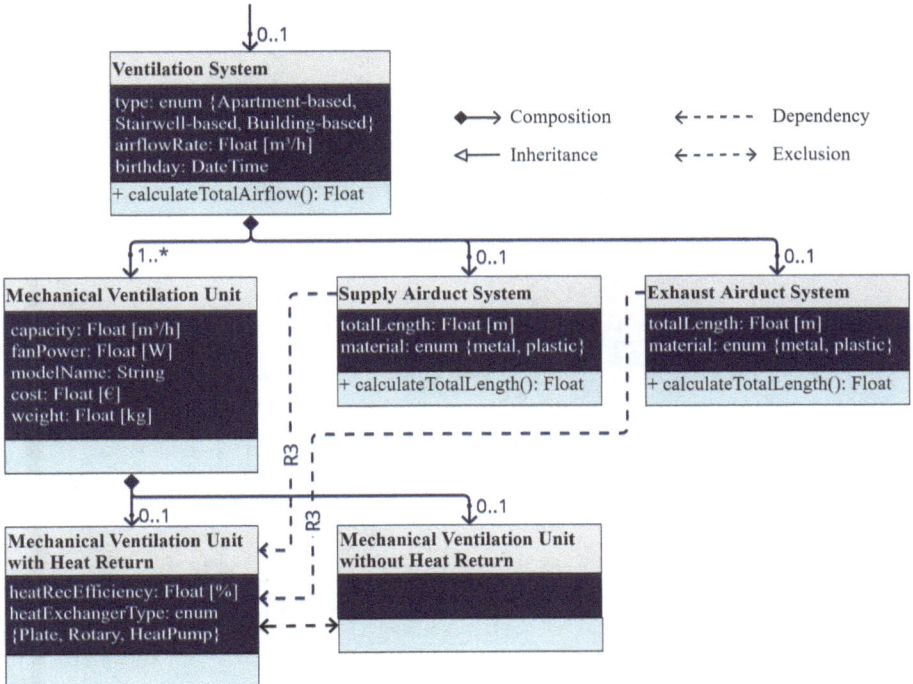

Fig. 1. Configuration Model rule R3 Unified Modeling Language (UML) schema example

The **interpretation module** is connected to the user interface, linking each input field value to the knowledge base. This module fulfils FRs no. 1, 2 and 3, where the system processes building information.

The **explanation module** is required to fulfil FR4 and for the system to communicate the advantages and disadvantages of different renovation choices.

The **solving module** fulfils FRs no. 5 to 7 through the following components:

1. Visualisation block shows a visual representation of the configuration. (FR no. 5 social and aesthetic requirements)
2. Compliance block identifies whether the configuration complies with public grant requirements and other important technical and legal requirements set for the building. (FR no. 5)
3. Energy Performance (EP) block assesses the energy performance (FR no. 6.b) and determines how the configuration complies with energy performance requirements. (FR no. 5)
4. GHG block calculates greenhouse gas emissions generated during building renovation. (FR no. 6.c)
5. Cost estimation block calculates the estimated construction costs and changes in residents' running expenses for the configuration. (FR no. 6.a)
6. Results block generates a comprehensive and sharable document describing the configuration (FR no. 7)

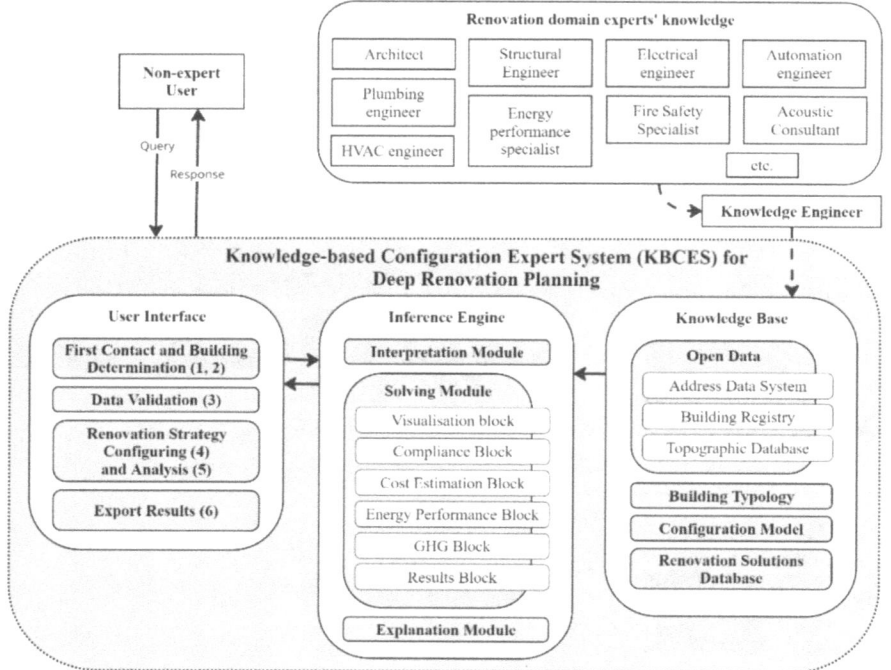

Fig. 2. System Architecture of the KBCES for Deep Renovation Planning

5.5 Proof-Of-Concept Prototype: Renokratt

In the DSR development stage, a web-based renovation KBCES proof-of-concept (PoC) prototype named Renokratt was created. The prototype's main purpose was to validate the functional requirements set previously. It was developed for Estonian Soviet-era apartment buildings, utilizing known typology knowledge in [8].

The prototype was developed on the **VIKTOR.ai** platform using **Python 3.11,** and the **PyCharm 2023.2.5** environment. **Rhino 7** and **Grasshopper** along with Rhino Compute were used to demonstrate possible workflows for the visualisation block of the solving module.

Building-specific knowledge was automatically retrieved from ADS, ETD and EBR using their respective application programming interface (API) services. The renovation measures database was created with MS Excel and the file was structured so that each Sheet corresponded to one component type (i.e., exterior wall, air leakage rate, heating source were on separate sheets). The inference engine used the Python Pandas package to structure the data into *DataFrame* objects for data analysis tasks such as enriching open data with typology knowledge and user input.

The EP block used the Estonian Renovation Strategy Tool (RST), which was originally designed to automate Urban Building Energy Model (UBEM) calculations for comprehensive renovation strategies at a neighbourhood-municipality scale [16]. This tool was effectively used to automate the preliminary energy performance calculation of a single building configuration.

The PoC prototype is accessible at [17], with information about ongoing development available at [18].

6 Demonstration

The prototype was qualitatively tested with four subjects in a lab environment. The testing process was structured into three stages:

1. **Context Introduction.** Renovation challenges faced by non-experts were validated. The system's purpose, user and objectives were described.
2. **Prototype Use.** The subjects were asked to use the prototype independently, assuming the role of a non-expert renovation initiator.
3. **Discussion.** The subjects shared feedback on the prototype's strengths, weaknesses, and potential areas for improvement.

The prototype Renokratt was developed to an extent that enabled us to demonstrate the use of functional requirements related to building identification and data verification (FR1–4). The implementation and demonstration of other features, such as compliance assessment (FR5) and financial and GHG impact assessments (FR6a, FR6c), require further development.

During testing, answers were sought for the following questions:

1. What are the biggest challenges related to initiating renovation? (Stage 1)
2. Studies have concluded that "the renovation initiator needs support in creating the design task". Do you agree with this statement and if yes, what kind of support is needed? (Stage 1)
3. How is the shown solution useful to you? (Stage 2)
4. Which KBCES components would help you most in your work? What should be prioritized during development? (Stage 3)
5. What are the most important parameters based on which the renovation client ultimately makes the renovation decision? (Stage 3)
6. Are the stated functional requirements appropriate to meet the objectives? (Stage 3)
7. How would such a tool help you in your daily work and renovation planning? (Stage 3)

Answers were found for all posed questions and valuable feedback was received for the created prototype and solution concept. Overall, all users agreed that the suggested concept would certainly be useful in their work processes if implemented to its full extent.

7 Evaluation

The development and testing of the KBCES concept and its proof-of-concept prototype Renokratt provided several insights into the digitalization of deep renovation planning. Testing validated both the relevance of renovation challenges identified in literature and the KBCES concepts suitability to address them.

The prototype Renokratt successfully demonstrated how integrating building typology data, public open data and expert knowledge can support setting preliminary renovation requirements for subsequent renovation stages. This aligns with the broader goals of increasing renovation efficiency through digitalization and expert knowledge formalization.

A significant insight from testing was the test subjects' emphasis on the renovation measures database and explanation module. The system's effectiveness depends heavily on implementing a comprehensive, filtered, and continuously updated database of easily understandable choices. This insight reinforces the need for structured, machine-readable documentation of renovation measures.

The system's effectiveness also depends on the quality of initial building data, and the user's ability to verify and enhance it. This data is crucial for accurately determining renovation scope and costs. This suggests that improved data structures in public and private building databases are a prerequisite for successful digitalization of renovation processes, aligning with the objectives set by the European Energy Performance of Buildings Directive (EPBD).

Additionally, the demonstration phase revealed that the most important decision-making factor for apartment associations, whether to undertake the renovation process or not, is financial impact–most notably the expected changes in monthly expenses for residents. Architectural and social aspects are secondary and energy performance improvements are often valued solely for their effect on real estate value rather than technical meaning.

The research contributes to the field by demonstrating how expert systems can bridge the knowledge gap between technical specialists and non-expert users in deep renovation planning. The validation of the KBCES concept provides a foundation for future development of digital renovation tools, particularly in contexts with similar building stock characteristics and renovation challenges.

8 Conclusion

The development of the Knowledge-Based Configuration Expert System (KBCES) for Deep Renovation Planning concept tackles key challenges in renovation planning and enables renovation initiators to articulate clear renovation requirements for design specialists. The proof-of-concept prototype Renokratt, developed for Estonian Soviet-era apartment buildings, demonstrated the system's suitability and relevance for this context.

A key strength of KBCES concept is its ability to reduce barriers for non-expert renovation initiators. The tool's architecture enables automatic building data collection, which, when validated and enriched with typology and user input, allows the generation of tailored renovation strategies.

Testing confirmed the technical feasibility of the system while highlighting the system's dependency on reliable input data. A comprehensive, up-to-date renovation measures database proved essential for successful implementation. Also, financial aspects, specifically changes in running expenses for residents, emerged as the primary decision-making factor for apartment associations.

The prototype demonstrates how digital tools can bridge the gap between technical expertise and end-user needs. With further development, Renokratt holds strong potential as a scalable tool supporting Estonia's renovation strategy and European sustainability objectives, in line with current trends in using digital technologies to meet sustainability and energy efficiency goals.

Funding. This work has been supported by the Estonian Centre of Excellence in Energy Efficiency, ENER (grant no. TK230), funded by the Estonian Ministry of Education and Research, and by personal research fundings "Data-Driven Governance Framework for Renovation Policy-Making, Decision-Making, and Management of Building Clusters" (grant no. PSG963). The study uses data from the project LIFE IP BUILDEST (LIFE20 IPC/EE/000010) funded by the European Commission.

References

1. MKM: Hoonete rekonstrueerimise pikaajaline strateegia. (2020). Accessed: 18 Sep 2023. https://kliimaministeerium.ee/elukeskkond-ringmajandus/elamud-ja-hooned/renoveerimis laine
2. Soonik, M., Roots, K., Roots, V., Viires, H.M.: Analüüs ja ettepanekud korterelamute renoveerimise protsesside tõhustamiseks, MKM, (2023). https://mkm.ee/ehitus-ja-elamumaja ndus/buildest/poliitikakujundamine
3. EKA and TalTech, 'Rohetiigri ehituse teekaart 2040', EKA, TalTech, Accessed: 18 Jan 2024 (2023). https://ehituseteekaart.rohetiiger.ee/
4. Kamari, A., Jensen, S., Christensen, M.L., Petersen, S., Kirkegaard, P.H.: A hybrid decision support system for generation of holistic renovation scenarios—cases of energy consumption, investment cost, and thermal indoor comfort. Sustainability **10**(4), 4, (2018). https://doi.org/10.3390/su10041255
5. Kamari, A., Schultz, C.P.L., Kirkegaard, P.H.: Constraint-based renovation design support through the renovation domain model. Autom. Constr. **104**, 265–280 (2019). https://doi.org/10.1016/j.autcon.2019.04.023
6. Peffers, K., Tuunanen, T., Rothenberger, M., Chatterjee, S.: A design science research methodology for information systems research. J. Manag. Inf. Syst. **24**, 45–77 (2007)
7. AL-Badareen, A., Selamat, M., Jabar, M.A., Din, J., Turaev, S.: Software quality models: a comparative study, presented at the Communications in Computer and Information Science. Jan 2011, pp. 46–55. https://doi.org/10.1007/978-3-642-22170-5_4
8. Iliste, E.: Ehitisregistri andmete alusel elamupiirkonna energiatõhususe hindamise alused. Accessed: 05 Sep 2023. https://digikogu.taltech.ee/et/item/db8623c0-7686-4c0e-9cca-c07 6543344a6
9. Arrak, K., Helilaid, M., Konov, V., Reiska, K., Schultz, A., Vaarik, R.: Ehitussektori digitaliseerituse uuring, Civitta Eesti AS, Lõõtsa **8**, 11415, Tallinn, Lõpparuanne (2024)
10. MKM: Korterelamute energiatõhususe toetuse tingimused. Accessed: 14 Feb (2025). https://www.riigiteataja.ee/akt/108032023013?leiaKehtiv
11. L., Chung, do Prado Leite, J.C.S.: On non-functional requirements in software engineering. In: Conceptual modeling: foundations and applications: essays in honor of john mylopoulos. Borgida, A.T., Chaudhri, V.K., Giorgini, P., Yu, E.S. (eds.) Berlin, Heidelberg: Springer, pp. 363–379 (2009). https://doi.org/10.1007/978-3-642-02463-4_19
12. Liao, S.-H.: Expert system methodologies and applications—a decade review from 1995 to 2004. Expert Syst. Appl. **28**(1), 93–103 (2005). https://doi.org/10.1016/j.eswa.2004.08.003
13. R. Akerkar and P. Sajja, *Knowledge-Based Systems.* Jones & Bartlett Learning, 2010
14. Felfernig, A., Hotz, L., Bagley, C., Tiihonen, J.: Knowledge-based configuration: from research to business cases. Newnes (2014)

15. Iliste, E., et al.: Heat loss characteristics of typology-based apartment building external walls for a digital twin-based renovation strategy tool. J. Phys. Conf. Ser. **2654**, no. (1), 012125 (2023). https://doi.org/10.1088/1742-6596/2654/1/012125

16. Hallik, J., Arumägi, E., Pikas, E., Kalamees, T.: Comparative assessment of simple and detailed energy performance models for urban energy modelling based on digital twin and statistical typology database for the renovation of existing building stock. Energy Build. **323**, 114775 (2024). https://doi.org/10.1016/j.enbuild.2024.114775

17. Viik, J.: Proof-of-concept Prototype Renokratt. Accessed: 01 Apr 2025. [Online]. Available: https://cloud.viktor.ai/public/renokratt

18. Renokratt, O.Ü.: Renokratt's homepage. Renokratt – Simplify and support your renovation projects. Accessed: 01 Apr 2025. [Online]. Available: https://www.renokratt.ee/

Evaluating the Adoption of XR Technologies for High-Risk Industrial Training Application

Caolan Plumb[1](✉), Farzad Rahimian[1], Diptangshu Pandit[1], Hannah Thomas[2], and Nigel Clark[2]

[1] Teesside University, Middlesbrough TS1 3BX, UK
caolan.plumb@faradaycentre.co.uk
[2] The Faraday Centre LTD, The Wilton Centre Annex, Middlesbrough TS10 4RF, UK

Abstract. Virtual Reality (VR) is increasingly recognised as a transformative tool for safety training, particularly in high-risk fields such as construction and electrical engineering. However, to successfully integrate this technology within established training practices, training outcomes ought to be rigorously validated. By investigating the effects of eXtended Reality (XR) technology upon the delivery of safety training for high-voltage operations, this paper proposes guidelines by which instructor adoption of XR tools may be assessed and seeks to establish a methodology to measure the effectiveness of training delivery. Factors influencing instructor adoption of XR tools and methods by which staff may be prepared to integrate new training methods are discussed. An exploration of relevant performance metrics, including trainee immersion and operational accuracy, informs the design of an experiment to compare the outcomes of different training sequences: participants are allocated into four experimental groups: (1) traditional training only, (2) VR training only, (3) traditional training followed by VR, and (4) VR training followed by traditional training. Immersive Tendencies Questionnaires (ITQs) assess trainee predispositions to immersion and Presence Questionnaires (PQs) assess perceived presence within the VTE. As well as detailing the collection of data, the experiment desgin also proposes statistical analyses to determine whether VR training complements traditional methods, enhances participant confidence, and improves knowledge retention.

Keywords: Virtual Reality Training · High-Voltage Safety · Training Validation · Immersion and Presence · Statistical Analysis

1 Introduction

1.1 Background and Motivation

Virtual Reality (VR) learning environments have emerged as powerful educational tools, particularly for high-risk industries requiring specialised equipment. Traditional training methods for the operation of high-voltage and extra high-voltage equipment involve practical difficulties such as operational down-time as live equipment is taken out of service to be made available training, or significant financial investment if procuring

© The Author(s) 2026
A. Jurelionis et al. (Eds.): BDTIC 2025, LNCE 775, pp. 155–165, 2026.
https://doi.org/10.1007/978-3-032-09040-9_14

expensive equipment for bespoke training purposes. Additionally, exposure to realistic hazards is limited by practical safety constraints. VR technology offers an alternative by providing trainees with safe, cost-effective, and customisable scenarios. The immersive and interactive nature of VR allows for realistic simulation of industrial systems, enabling trainees to develop knowledge and confidence without the risks or operational costs associated with actual equipment handling [1–4].

Due to the role instructors play in guiding and assessing trainee learning outcomes, successful VR implementation depends not only on the development of robust software but also on instructor proficiency and comfort. Thus, we are motivated to ensure that instructors are accepting of and confident with XR training tools. Secondly, to establish the effectiveness of VR experiences for trainees, we are motivated to propose a clear methodology for the validation of training outcomes, thereby aiding integration of VR technology for industrial safety training.

1.2   Project Context: Faraday Nexus at the Faraday Centre

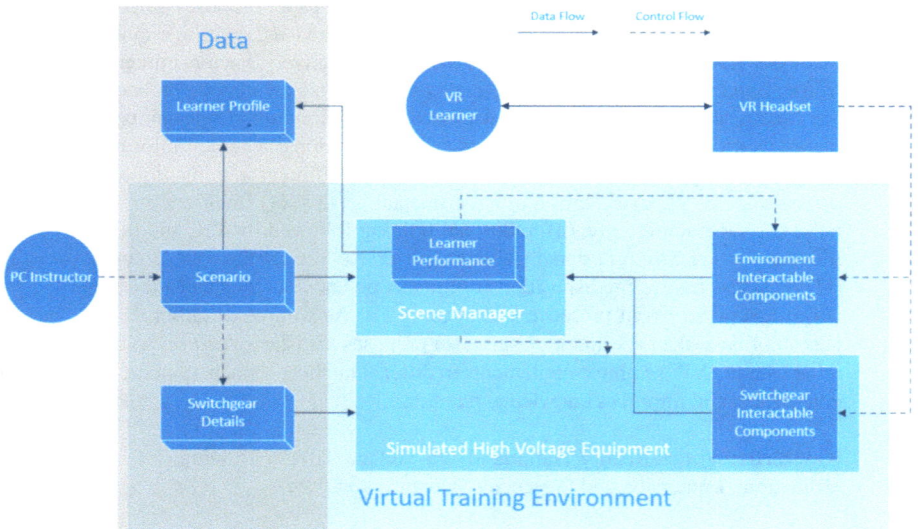

Fig. 1. Overview of the structure of the Faraday Nexus VTE

Unreal Engine has been used to develop a virtual training environment (VTE) for The Faraday Centre LTD under the project title "Faraday Nexus." Within this VTE, VR users (trainees) are presented with interactive reconstructions of high-voltage and extra-high-voltage equipment. The Faraday Nexus expects the connection of a second device to the training session, operated by an educator or instructor. This second user connects to VR user via a PC or laptop to control and supervise the experience of the trainee(s) as they are immersed in the VTE. Instructors set scenario parameters which author the sequence of actions expected to be performed by the trainees; parameters including equipment type, hazard presence, and simulated malfunctions (Fig. 1).

The instructor spectates the trainee's VR experience and offers guidance by explaining necessary tasks, or by navigating the PC application to highlight relevant pieces of equipment for the trainee. By simulating the operation of real-world equipment, the Faraday Nexus doubles as both training platform and assessment tool. When the platform is being used to assess trainee performance, instructors are still able to provide guidance on the specifics of equipment operation, but trainees are expected to autonomously adhere to the safety principles governing the switching scenario. The VTE has been designed to recognise any major or minor mistakes in the trainee's execution of the switching procedure and measures the accuracy of the trainee's performance the way an instructor might.

1.3 Goals of this Article

As the Nexus project progresses towards commercial implementation, ensuring instructor proficiency with and acceptance of XR training tools becomes a priority, alongside assessments of training outcomes relative to traditional methods. The primary goal of this article is to provide a framework for improving instructor adoption of VR-based training and to validate the effectiveness of this training compared to traditional methods. It is hoped that this framework will later be employed to confirm the reliability and educational value of VTEs.

This paper is structured as follows: Sect. 2 reviews the role played by instructors in the adoption of XR technologies for commercial training in high-risk sectors, specifically electrical engineering. Section 3 identifies metrics valuable to the analysis of trainee performance when receiving VR training. Section 4 proposes a framework which may be employed to validate the effectiveness of VR training and investigate potential synergies with existing training methods. Section 5 concludes the paper with a review of the contributions as well as the limitations of the proposed framework.

2 Evaluating Instructor Adoption of VR Technologies

2.1 The Importance of Instructor Adoption

VR remains relatively novel for many educators, particularly those without prior experience with immersive technologies. Anecdotal observations of our case study at The Faraday Centre LTD suggest that many electrical engineering lecturers, despite extensive professional and pedagogical expertise, are not yet fully literate with VR technology. Instructor unfamiliarity may lead to error, hesitation, or improper utilisation, diminishing the effectiveness of VR-based training. The success of the commercial implementation of the VTE will rely not only on the development of robust software, but also effectively preparing instructors to mediate with the technology. Therefore, instructor training and familiarisation represent important preliminary steps toward effective deployment of virtual training environments for safety training purposes.

2.2 Key Areas for Investigation and Instructor Feedback

Immersive Tendencies: Immersive Tendency Questionnaires (ITQ) have proven valuable in understanding user predisposition toward immersive experiences and can predict instructor acceptance and response to VTE interfaces [5, 6].

Usability: Usability refers to the measurement of how easily a user can accomplish their goals when using a service. This is usually measured through usability questionnaires[7]. Quantitative usability scales will be combined with qualitative interview data to provide insights into instructor acceptance and possible usability issues [8, 9].

Simulation Fidelity and Accuracy: Prior research emphasises the critical importance of validating the perceived fidelity of the simulation from domain experts' perspectives. It is particularly vital in safety-critical environments, ensuring instructors view the simulation as realistic, relevant, and trustworthy [10, 11].

Control Interfaces and Interaction Design: Including customisation options for varied user demographics, such as increasing font size for the visually impaired or colourblind settings would improve project accessibility. Studies underline the importance of intuitive interaction design and customisable control schemes to facilitate adoption among diverse user groups, especially those less familiar with VR technology [12, 13].

2.3 Instructor Onboarding and Familiarisation

The process of ensuring instructor proficiency and confidence with the Faraday Nexus will begin with an initial orientation workshop, introducing the VTE's objectives, navigation controls, and instructor interface. Instructors will receive a demonstration of a full training scenario, including instructor monitoring tools.

Following this, instructors will undergo guided hands-on sessions using the system in mock training scenarios, allowing them to experience the VR interface from a trainee perspective. After being tasked with setting up and running a training session themselves using scenario configuration tools, opportunities will be provided for instructors to reflect and discuss usability and control scheme intuitivity. Post-training surveys (e.g. System Usability Scale) and structured interviews will assess instructor comfort, perceived effectiveness, and suggestions for refinement. Instructors will also be asked for feedback on the accuracy of their experience within the VTE and to evaluate the realism of the training scenario compared to industrial scenarios.

Instructors will receive ongoing support including access to technical assistance, updated digital manuals, and a channel of communication with the development team for feature improvements, encouraging the staff to play an active role in the deployment of VR training (Fig. 2).

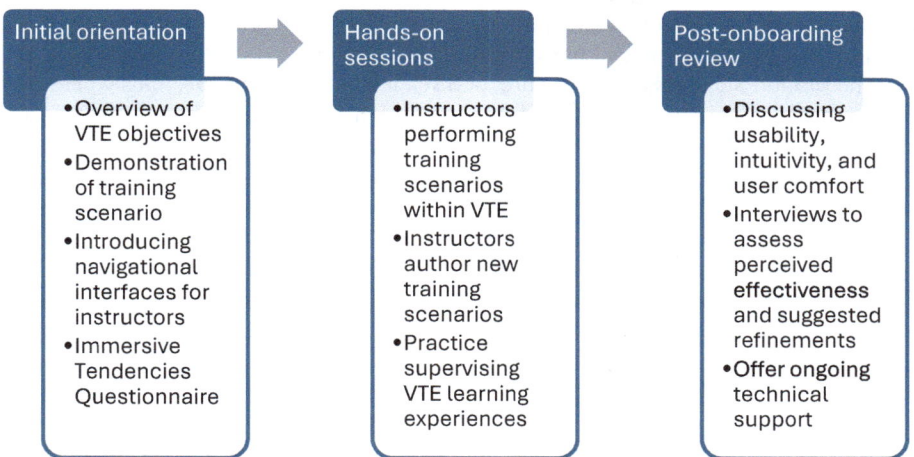

Fig. 2. Overview of the proposed instructor onboarding process

3   Evaluating Effectiveness of VR Training Environments

3.1   Quantifying Training Effectiveness

Procedural Accuracy: Measuring how correctly trainees execute switchgear operations by counting the number of mistakes. Over multiple training scenarios, if trainees can consistently replicate procedures without errors, then it is likely that the trainee has achieved a standard of competence necessary for real world application [14, 15].

Response Time: We will measure the time it takes the trainee to complete the training procedure. By gathering this information from the tests, we can compare this metric with the trainees' qualititative surveys to evaluate if there is indeed a correlation between fast, accurate responses and increased levels of trainee confidence [16, 17].

Confidence Levels: Measuring confidence (via self-reports or behavioral observations) shows whether trainees feel ready and sufficiently trained to act correctly under pressure [18, 19].

3.2   Factors Which Influence User Experience in VR

Like instructor adoption validation, a pre-assessment phase involving an ITQs will help understand trainees' dispositions towards the VTE as a learning device. Once users have performed their training, there will be several similar qualitative surveys they will be asked to fill in to measure their experience within VR.

Presence is a highly important factor in measuring VR experiences, especially for training and education purposes [5]. The Presence Questionnaire, or PQ [20], is one of the most commonly used questionnaires in VR research [21] and consists items divided into five domains ("realism," "possibility to act," "quality of inter-face," "possibility to examine," and "self-evaluation of performance"). The PQ has shown to be robust and reliable to analyse the sense of presence [20].

The Motion Sickness Questionnaire, or SSQ, is a collection of questions where each relates to a symptom of motion sickness—nausea, oculomotor, and disorientation—and trainees should answer according to the severity of each symptom ranging from "none" to "severe." The total score of this questionnaire provides an insight into motion sickness severity [22, 23].

The Faraday Nexus could be further gamified to see if this influences user experience or immersion within the training simulation. Further gamification options may include positive reinforcement of correct interaction via a scoring system, additional visual effects, or including achievements. These features could be activated and deactivated individually to control later tests for any effect on retention of trainee focus.

4   Experiment Design

4.1   Experiment Objectives

To enable a comparative evaluation of training outcomes across different approaches to training delivery, participants will be randomly assigned to one of four groups: Traditional only, VR only, and two hybrid approaches combining VR and traditional in different orders. This four-group design has been selected to investigate whether incorporating mixed-reality technology with traditional methods leads to improved training outcomes and whether the sequence of training delivery influences effectiveness. This investigation can be summarised by the following three primary hypotheses:

1. VR training alone will be at least as effective as traditional training.
2. Hybrid training (Traditional $\to$ VR or VR $\to$ Traditional) will outperform single-method training in knowledge retention and confidence.
3. The order of training (VR first vs. Traditional first) may impact learning effectiveness.

Additionally, information gathered during this experiment may also be used to investigate secondary correlations, which do not reflect on the validity of VR/hybrid training but give insight into the experience of trainees' experiences in VR. Higher immersive tendencies (ITQ) are expected to predict greater perceived presence (PQ), which in turn may be associated with (a) greater improvements in confidence, (b) higher procedural accuracy, (c) faster response times, and (d) reduced simulator sickness (SSQ). These secondary hypotheses explore the extent to which trainee disposition and perceived presence correlate with the overall effectiveness of the VR training experience (Table 1).

Table 1. Outlining the four groups for experiment

Group	Training Method	Purpose
G1: Traditional Only (Control Group)	Traditional training only	Baseline for comparison

(*continued*)

Table 1. (*continued*)

Group	Training Method	Purpose
G2: VR Only (Experimental Group)	VR training only	Tests VR as a standalone method
G3: Traditional → VR (Hybrid Group 1)	Traditional training first, then VR	Tests if VR reinforces traditional methods
G4: VR → Traditional (Hybrid Group 2)	VR training first, then traditional	Tests if traditional training reinforces VR learning

4.2 Data Collection Plan

Each participant in the study will complete a structured sequence of assessments designed to evaluate both training effectiveness and experiential factors. Prior to training, all participants will complete a self-reported confidence survey using a 1–5 Likert scale. Those in VR-involved groups (G2, G3, G4) will also complete the Immersive Tendencies Questionnaire (ITQ) to measure their predisposition toward immersive technology. Participants will then receive training based on their assigned group: G1 will undergo traditional switchgear training only; G2 will complete training entirely in VR; G3 will first complete traditional training followed by VR; and G4 will receive VR training followed by traditional instruction.

Immediately following the training session, all participants will complete a procedural accuracy test to assess their ability to perform switching procedures, and their response time will be recorded. A post-training confidence survey will be conducted to evaluate perceived learning gains. The procedural accuracy test and confidence reports will be used to quanitfy and compare the results of the different experiment groups. VR participants (G2, G3, G4) will also complete the Presence Questionnaire (PQ) to measure their sense of immersion, the System Usability Scale (SUS) to evaluate interface usability, and the Simulator Sickness Questionnaire (SSQ) to measure any adverse physical effects experienced during VR exposure.

Once a training session has been concluded, instructors will be asked to evaluate the ease of teaching their group (G1–4), and record how enthusiastic participants appeared. The opportunity for instructors to provide feedback on the session will allow for more usability comments to be made, providing greater insight into potential feature improvements or future areas for VTE development. Similarly, trainees will be asked to complete a qualititave survey of their experience and may suggest improvements.

Approximately one to two weeks later, participants will return for a delayed retention test. This session includes a second procedural accuracy assessment to evaluate retained competence, a knowledge retention questionnaire focused on switching procedure understanding, and a final confidence survey to assess longer-term perceptions of preparedness. Experience (Table 2).

Table 2. Pre/post training data collection

Metric	Collected from	Pre-Training	Post-Training?	Delayed Test?
Confidence (1-5 Scale)	All groups			
Procedural Accuracy	All groups			
Response Time	All groups			
Immersive Tendencies Score (ITQ)	VR Groups Only (G2, G3, G4)			
Presence Score (PQ)	VR Groups Only			
Usability Score (SUS)	VR Groups Only			
Severity of Motion Sickness (SSQ)	VR Groups Only			

Instructor Evaluation: Instructors assess ease of learning and engagement.

Trainee Feedback: Structured survey on perceived usability suggested refinements.

4.3   Statistical Analysis Plan

To evaluate overall training effectiveness and potential differences arising from the sequence of training methods, several statistal tests have been chosen to analyse quantitive data gathered from the experiment:

Independent t-tests will be used for direct comparisons between two groups to identify differences in performance metrics (assuming normal distribution). For example, when comparing the procedural accuracy between the baseline group (G1: traditional training methods only) and the VR-only group (G2). If data is not normally distributed, or if the sample size is too small, a Mann-Whitney U Test will be employed as a non-parametric alternative to compare results between specific groups.

Two-Way ANOVA is employed when comparing multiple training groups to determine if hybrid approaches improve training outcomes. This test is also capable of analysing any effects of training sequence, seeking to clarify if training order plays a role.

The Repeated Measures ANOVA is applied in this context to validate whether training effects are sustained over time and will allow for comparison of retention curves between training methods. Specifically, is post-training confidence greater than pre-training confidence, is confidence tested 1–2 weeks later greater than pre-training confidence, and is confidence tested 1–2 weeks later roughly equal to post-training confidence (inferring confidence retention).

Correlations between trainees' self-reported dispositions towards immersive technologies, their subjective experiences within the VTE, and their performance metrics will be evaluated using analyses like Pearson correlation (if both values in question are normally distributed) or Spearman correlation (if values are not normally distributed). These correlation analyses will provide insight into the influence of presence on the outcomes of VR and hybrid training, and whether a trainee's disposition towards immersive technologies may predict more effective training (Table 3).

Table 3. Statistical analysis plan

Analysis Description	Comparison	Statistical Test
Effectiveness of VR vs. Traditional	G1 vs. G2	Independent Samples t-Test (or Mann-Whitney U if data isn't normally distributed)
Does order matter in hybrid training?	G3 vs. G4	Independent Samples t-Test (or Mann-Whitney U if data isn't normally distributed)
Do hybrid methods outperform single methods, and does the order of hybrid training delivery matter overall?	(G1 & G2) vs. (G3 & G4)	Two-Way ANOVA
Confidence retention over time (Pre-Test, Post-Test, Delayed Test)	All Groups	Repeated Measures ANOVA
Correlations between individual trainee dispositions and subjective experience		
Analysis Description	**Comparison**	**Statistical Test**
Do higher immersive tendencies predict higher perceived presence?	ITQ vs. PQ	Pearson/Spearman Correlation
Does disposition towards immersive technologies affect training outcomes?	ITQ vs. SSQ, Accuracy, Change in confidence, Response time.	Pearson/Spearman Correlation
Does perceived presence affect training outcomes?	PQ vs. SSQ, Accuracy, Change in confidence, Response time	Pearson/Spearman Correlation

5   Conclusion and Areas for Further Investigation

It is hoped that the adoption guidelines provided will facilitate access to VTEs for educational institutions. Improved instructor proficiency and confidence is expected to smooth the process of incorporting XR tools to supplement traditional training methods. Validating and optimising VR-based training will help lead to safer, more cost-effective, and scalable educational solutions for high-risk environments.

By comparing VR, traditional, and hybrid training methods across procedural accuracy, response time, confidence, and knowledge retention, the designed framework for experimentation aims to clearly identify strengths and limitations of each approach. By preparing to evaluate training outcomes, this research supports the adoption and validation of the effectiveness of VTEs, specifically for HV electrical engineering.

While the outlined methodology readies investigators to receive some valuable insights, it also presents limitations. Findings from a single training context may not generalise across all high-risk industries and the study's scope for long-term evaluation of training impacts is limited to two weeks.

Future research should consider broader demographic diversity, such as comparing the effects of introducing VR training for different industries. Later, the opportunity to perform further delayed retention tests to collect data over a longer period which may provide greater insights into training effectiveness. Lastly, if the outcomes of the proposed experiment indicate that heightened perceived presence correlates positively with trainee performance metrics, further investigation into the effects of greater gamification within the VTE (alongside other methods of potentially enhancing presence) may provide novel techniques for improving training outcomes.

References

1. Adami, P., et al.: Effectiveness of VR-based training on improving construction workers' knowledge, skills, and safety behavior in robotic teleoperation. Adv. Eng. Inform. **50**, 101431 (2021)
2. Stefan, H., et al.: Evaluating the preliminary effectiveness of industrial virtual reality safety training for ozone generator isolation procedure. Saf. Sci. **163**, 106125 (2023)
3. White, W.W., Jung, M.J.: Three-Dimensional Virtual Reality Spinal Cord Stimulator Training Improves Trainee Procedural Confidence and Performance. Neuromodulation: Technology at the Neural Interface (2022)
4. Plumb, C., et al.: A framework for realistic virtual representation for immersive training environments. In: Proceedings e report, pp. 274–287. Firenze University Press (2023)
5. Grassini, S., Laumann, K., Rasmussen Skogstad, M.: The use of virtual reality alone does not promote training performance (but sense of presence does). Front. Psychol. **11** (2020)
6. Witmer, B.G., Singer, M.J.: Measuring presence in virtual environments: a presence questionnaire. presence: Teleoperat. Virtual Environ. **7**(3), 225–240 (1998)
7. Brooke, J.: SUS-A quick and dirty usability scale. Usability Eval. Indus. **189**(194), 4–7 (1996)
8. Papachristos, N.M., Vrellis, I., Mikropoulos, T.A.: A comparison between oculus rift and a low-cost smartphone VR headset: immersive user experience and learning. In: 2017 IEEE 17th International Conference on Advanced Learning Technologies (ICALT) (2017)
9. Radianti, J., et al.: A systematic review of immersive virtual reality applications for higher education: design elements, lessons learned, and research agenda. Comput. Educ. **147**, 103778 (2020)
10. Dieckmann, P., Gaba, D., Rall, M.: Deepening the theoretical foundations of patient simulation as social practice. Simul. Healthc. **2**(3), 183–193 (2007)
11. Alhalabi, W.: Virtual reality systems enhance students' achievements in engineering education. Behav. Inform. Technol. **35**(11), 919–925 (2016)
12. Jerald, J.: The VR Book: Human-Centered Design for Virtual Reality. Morgan & Claypool (2015)

13. Scerbo, M.W., et al.: The efficacy of a medical virtual reality simulator for training phlebotomy. Hum. Factors **48**(1), 72–84 (2006)
14. De Lorenzis, F., et al.: Immersive Virtual Reality for procedural training: comparing traditional and learning by teaching approaches. Comput. Ind. **144**, 103785 (2023)
15. Bloom, M.B., et al.: Virtual reality applied to procedural testing: the next era. Ann. Surg. **237**(3), 442–448 (2003)
16. Hawes, J., Ward, G.: The Impact of Training Reaction Times and Accuracy with Motion Controllers in Virtual Reality (2024)
17. Casella, A., et al.: Effects of a virtual reality reaction training protocol on physical and cognitive skills of young adults and their neural correlates: a randomized controlled trial study. Brain Sci. **14**(7) (2024)
18. Best, P., et al.: Immersive virtual environments as a tool to improve confidence and role expectancy in prospective social work students: a proof-of-concept study. Social Work Education, pp. 1–19
19. Liaw, S.Y., et al.: Assessment for simulation learning outcomes: a comparison of knowledge and self-reported confidence with observed clinical performance. Nurse Educ. Today **32**(6), e35–e39 (2012)
20. Witmer, B.G., Jerome, C.J., Singer, M.J.: The factor structure of the presence questionnaire. Presence: Teleoperat. Virtual Environ. **14**(3), 298–312 (2005)
21. Hein, D., C. Mai, 2005, Hußmann, H.: The usage of presence measurements in research: a review. In: Proceedings of the International Society for Presence Research Annual Conference (Presence). The International Society for Presence Research Prague (2018)
22. Balk, S.A., Bertola, D.B., Inman, V.W.: Simulator sickness questionnaire: twenty years later. In: Driving Assessment Conference. University of Iowa (2013)
23. Kennedy, R.S., et al.: Simulator sickness questionnaire: an enhanced method for quantifying simulator sickness. Int. J. Aviat. Psychol. **3**(3), 203–220 (1993)

DCU Campus Explorer: Engaging Staff and Students to Use Campus Facilities

Jaime B. Fernandez[1,2](✉) ⬤, Darragh Nagle[1] ⬤, Annabella Stover[1] ⬤,
Fiona Earley[1] ⬤, Kieran Mahon[1] ⬤, Tomas E. Ward[1,2] ⬤, Noel E. O'Connor[1,2] ⬤,
and Muhammad Intizar Ali[1,2] ⬤

[1] Dublin City University, Dublin, Ireland
{jaimeboanerjes.fernandezroblero,Annabella.stover,Fiona.earley,
kieran.mahon,tomas.ward,noel.oconnor,ali.intizar}@dcu.ie,
darragh.nagle4@mail.dcu.ie
[2] Insight Research Ireland Centre for Data Analytics, Dublin, Ireland

Abstract. CE (Citizen engagement) is an activity that aims to provide a way to link citizens and organizations. As such it has a bidirectional channel of communication, from citizens to organization and from organization to citizens. These channels of communication should be user friendly, easily accessible and also need to fulfil other requisites depending on the use case. In that sense, CE still poses a research challenge to provide a complete solution. To this end, Digital Twin (DT) technology has been used to develop tools that integrate and visualize data of different types like 2D/3D maps digital models, sensor readings and any multimedia data. DTs have the capacity to provide as well user-friendly and easily accessible interfaces. This work focuses on the application of DT technology to DCU (Dublin City University), specifically the DCU C&C (Care & Connect) initiative that was charged with raising awareness of the services, facilities, and increasing feeling of belonging and inclusion across the DCU campuses. Services such as sexual health services, wellness support, and facilities like quiet rooms, sensory pods, and room locations. The use of DT technology has enabled the development of a live Campus Explorer with which students can interact in a user-friendly and easily accessible manner on a web-based interface using mobile phones, tablets or laptops. This paper provides an overview of the DCU Campus Explorer and how it is being used in practice by 22,000 students, more than 3,000 staff plus visitors at https://www.dcu.ie/CampusExplorer.

Keywords: Digital Twins · Smart Universities · Citizen Engagement · Virtual Environments · Virtual Reality

1 Introduction

Citizen engagement, often seen as a two-way dialog between citizens and the government, is an important activity in which the citizen is a stakeholder, or a person likely to be impacted [1]. Citizens play a critical role in advocating transparency, holding public institutions accountable and contributing to their effectiveness. They also provide

A. Jurelionis et al. (Eds.): BDTIC 2025, LNCE 775, pp. 166–178, 2026.
https://doi.org/10.1007/978-3-032-09040-9_15

innovative solutions to complex development challenges. Growing evidence suggests that, under the right conditions, meaningful forms of citizen engagement (CE) can lead to better governance, citizen empowerment, more constructive citizen-state relations, strengthened public service delivery, and ultimately enhanced development effectiveness and well-being. Social, political, economic, environmental, cultural, geographic, and gender dynamics all shape the opportunities and scope for effective citizen engagement. Understanding the context in which CE practices occur and supporting enabling conditions are essential for achieving results. However, the outcomes of citizen engagement are highly context-specific and depend on the capacity and willingness of both governments and citizens to engage. In recent years, global research and practice have placed greater emphasis on making citizen engagement and social accountability practices more strategic and effective [2]. To this end, several studies have been carried out on CE [3] and in recent years several of them have evidenced the role of digital tools on this [4–7].

Universities are part of the institutions that want to improve their services and facilities based on students and staff engagement. For this purpose, they are developing their own tools that fit their own needs. One of the emerging technologies that are being used is DTs (Digital Twins). This technology has the capacity to integrate geospatial information, environmental sensor readings and any multimedia data. All this on top of user-friendly and accessible interfaces. Two features that are key for having a real impact in the CE activity. However, DT technology is still to be explored as there is not a complete solution that fulfils all the requirements of CE in real-life scenarios. This work presents a methodology based on DTs to develop a Campus Explorer for DCU (Dublin City University) that allows DCU community to explore DCU campuses in an immersive, user-friendly and easily accessible interface.

This paper is structured into six sections – Sect. 2 covers the literature review related to the use of digital twin in universities. Section 3 describes DCU Campus Explorer objectives and requirements. Section 4 discusses the methods and materials used for this work. Section 5 covers the results and findings. Finally, the conclusions, limitations and future research directions are discussed in Sect. 6.

2 Related Work

Considering the significant interest in CE (Citizen Engagement) and DTs (Digital Twins) across a range of organizations and institutions, several universities around the world are implementing their CE tools to visualize different types of information related to their campus. Kaunas University of Technology (Lithuania) applies DT technology at campus level to estimate operational carbon emissions, monitor indoor climate, and improve the energy performance of university buildings. The system integrates a geometrical 3D model with sensor data, supporting the monitoring of the university environment. APIs broadcast more than 2,000 parameters in real time from physically installed sensors across the university campus, with further data processing and analysis [8]. Another example is Western Sydney University (Australia) which explores the use of Digital Twin technology within the concept of a "living lab" for a university library building. This approach supports building facility management by proactively optimizing indoor conditions and simultaneously considering occupancy levels based on motion detection [9]. The University of Galway (Ireland) has developed a three-dimensional virtual model of the campus, created using a combination of surveys, RGB images, and photogrammetry. This model supports master planning procedures, provides a visualization of the university environment, and aids in planning security and traffic [10]. A similar approach is adopted by the University of Birmingham (United Kingdom) which uses 360-degree views and photographs of various points of interest, including campus buildings. Additionally, the university has outlined a strategy for further developing its smart campus, which includes an integrated BIM strategy, occupancy monitoring, and a navigation and wayfinding platform [11]. Hubei University of Technology (China) also employs a smart campus system based on DTs. The architecture of the system combines a static virtual model with a dynamic data model. It is primarily used for visualizing a virtual campus environment, offering features such as campus guided tours, augmented reality experiences, and virtual simulation teaching [12]. Recently, more complex use cases have been done, such as in [13] where campus capacity analysis and optimization are targeted, and in [14, 15] where their application besides including campus facilities information, they are also using their DTs to improve mobility, sustainability and accessibility in their university. The present overview indicates that DT technology has the potential to facilitate the development of smart university applications such as CE tools. It can be used to improve a wide range of services and support that universities can offer to their students, which is a crucial aspect of improving university services and facilities. Looking at the successful applications of DT technology in other universities, this research aims to explore its application in a real-life scenario in Dublin City University specifically for citizen engagement purposes.

3 DCU Care and Connect and Beyond

DCU C&C (Care & Connect) is a whole-of-university approach that situates national frameworks and initiatives under an overarching DCU Health and Wellbeing Strategy. It aims to develop an inclusive and welcoming campus environment, with departments across the university contributing to the delivery of a quality campus experience. The DCU C&C team works alongside DCU people to support the delivery of quality campus experience for staff, students and visitors.

Currently, although DCU units provide different services and facilities around the campuses to students, staff, and visitors, often these stakeholders do not know about them, or in some cases they struggle to find them. To address this, the aim is to create a campus engagement tool named Campus Explorer, that facilitates the communication and dissemination of formal and informal relevant information to the DCU community.

Insight (Insight Research Ireland Centre for Data Analytics) at DCU (Dublin City University) is a research centre with a great expertise in data analysis and AI applications. Currently, Insight is working on the development of the DT (Digital Twin) of DCU. The objective of this project is to create a DT that in the future can fit different purposes and use cases. After four years of development, this project has reached a certain maturity, therefore it is now looking for impactful use cases where the developed DT can be deployed. Collaborating with DCU C&C to create the DCU Campus Explorer was an ideal opportunity in this regard.

Along with the presentation of the current DT technology to DCU, the task of identifying some requirements for the Campus Explorer was carried out by engaging directly with DCU C&C units so they could express their need to us. Once these needs were collected the most relevant were extracted. This activity was important to determine if the current digital twin technology fits this real-life use case. The following important requirements were identified:

1. **Quick overview of the DCU campuses:** the first requirement was that the Campus Explorer allows quick overview of the important areas of the three teaching campuses. This way the users can familiarize themselves quickly with the areas without the need to overlay any information.
2. **Building information:** the next requirement was to facilitate the presentation of information about buildings such as location and entrances.
3. **Services and facilities:** the Campus Explorer should provide information of the different services available such as sexual health services, wellness support, and facilities such as quiet spaces, spaces to study, and universal access bathrooms.
4. **Outdoor path finding:** people normally struggle to reach buildings and areas in DCU therefore some predefined routes to specific areas or buildings of interest are needed.
5. **Tours:** DCU runs open days and orientation periods in DCU for new students and staff. In that sense, they wanted the functionality to create some interactive tours on the Campus Explorer.
6. **Update of the new renovated areas:** DCU Glasnevin campus just built a high-tech building called the Polaris. The requirement was that this Campus Explorer has the capacity to include updates of renovated areas.

4   Methods and Materials

After analyzing the requirements from DCU C&C, the methodology shown in Fig. 1 was developed to fit most of the requirements. This methodology is an extension of the previous works presented in [16, 17]. While in the previous works we cover a general methodology to create a digital twin, such as the creation and the integration of digital assets, and the streaming of IoT sensors. This use case dictates its own requirements such as the proper organization of layers of information. Information that, different to IoT sensor data, does not need to be updated in real time, but it must be better structured and accompanied by multimedia data. Also, in this case, because of the limited time and resources available to have a working Campus Explorer released on certain date, to show visual information of some facilities instead of creating 3D digital models of them, 360 images and videos of such places were collected. The integration of this multimedia data is also presented in the current methodology.

4.1   Data Acquisition

Once the requirements from DCU C&C department were identified, the next activity was the collection of information about the campus. A lot of focus was put into this activity since this information is to be overlayed on top of the digital twin. At this point, we only need to collect and write down all the information that is planned to be shared

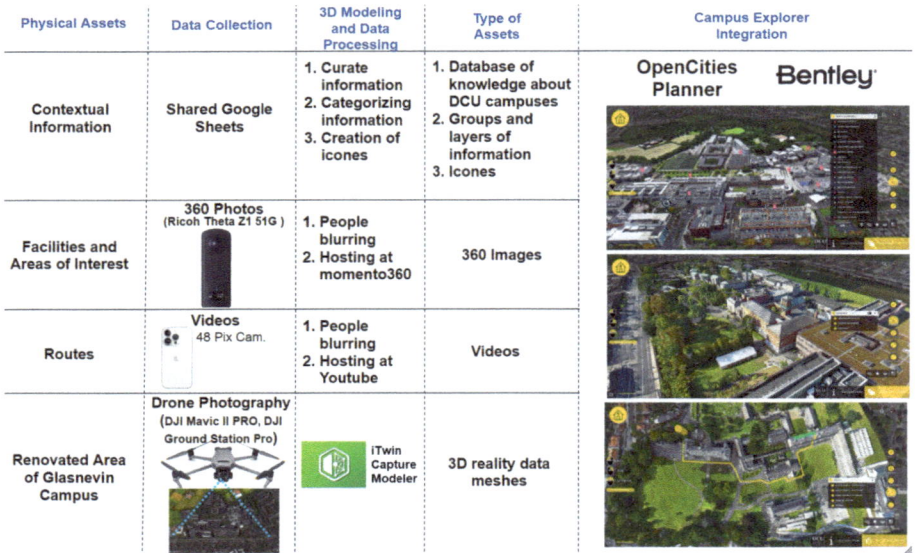

Physical Assets	Data Collection	3D Modeling and Data Processing	Type of Assets	Campus Explorer Integration
Contextual Information	Shared Google Sheets	1. Curate information 2. Categorizing information 3. Creation of icones	1. Database of knowledge about DCU campuses 2. Groups and layers of information 3. Icones	OpenCities Planner Bentley
Facilities and Areas of Interest	360 Photos (Ricoh Theta Z1 51G)	1. People blurring 2. Hosting at momento360	360 Images	
Routes	Videos 48 Pix Cam.	1. People blurring 2. Hosting at Youtube	Videos	
Renovated Area of Glasnevin Campus	Drone Photography (DJI Mavic II PRO, DJI Ground Station Pro)	iTwin Capture Modeler	3D reality data meshes	

Fig. 1. DCU Campus Explorer methodology.

with the DCU community and visitors. Therefore, a document was created and shared with the corresponding staff of DCU C&C to include all the information that they think is relevant.

From previous experience, we know all the work and resources that are needed to do indoor 3D scanning. Therefore, because of the limited resources and time available for this project, no new indoor 3D mapping was done. Instead, to cover more spaces and still give a sense of 3D views, 360 images were collected using a 360-camera model Ricoh Theta Z1 51G. To take these pictures a tripod was needed to place the camera which is then controlled remotely using a phone.

Videos were captured using an iPhone 14 Pro Max. Most of these videos were collected to show routes from the main entrance of DCU to buildings or areas of interest such as the main reception, library, and canteens. The idea is that these videos are posted along with a visual path overlayed on top of the 3D meshes of the campus.

On the other hand, a new drone survey was executed on an area of DCU Glasnevin campus where a new high-tech building was built, and it needed to be included in this digital twin. The technique and equipment used for this survey is the same as present in [16, 17], but for a smaller area.

4.2 3D Data Modeling and Processing

Once the raw information was collected, DCU C&C staff took the task of organizing such information in groups and layers so it can be easily understood and accessed by the final users. This organization of information is key in this work, since it sets the layout of the DCU Campus Explorer. It represents the expertise, needs, and objectives of the DCU units. This way, the developed Campus Explorer will really impact on the

DCU community. The organized information is kept on excel sheets where all changes and modifications can be tracked, Fig. 2. These excel sheets constitute our database of knowledge that later will be translated into groups, layers and POIs in the digital twin.

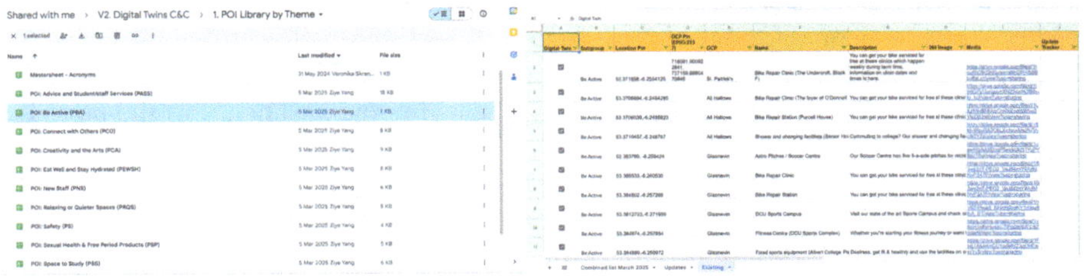

Fig. 2. Organization of contextual information.

Along with the organizing of contextual information, another important activity carried out was the creation of icons specifically for DCU, therefore from this task a collection of icons and its context was created, as presented in Fig. 3.

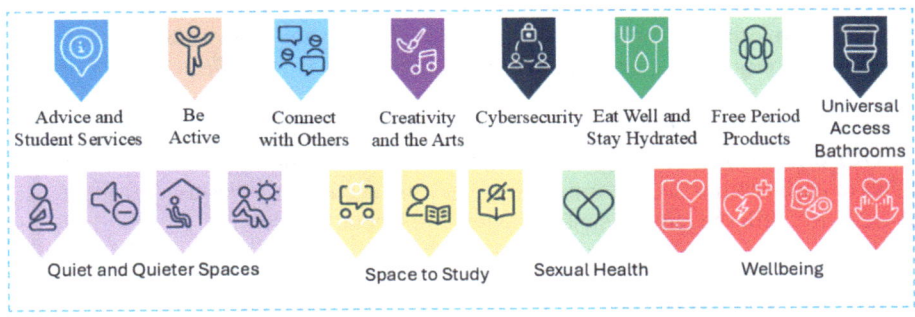

| Advice and Student Services | Be Active | Connect with Others | Creativity and the Arts | Cybersecurity | Eat Well and Stay Hydrated | Free Period Products | Universal Access Bathrooms |

Quiet and Quieter Spaces   Space to Study   Sexual Health   Wellbeing

Fig. 3. DCU icons representing different types of services and facilities.

During the collection of 360 images, it was inevitable that people appeared on them. Therefore, to comply with GDPR requirements all the areas of the images where a person appears were blurred so this person cannot be recognized. After that, the 360 images were hosted at momento360. This tool allows to host the images on the cloud. For each image a public URL link is created which will permit us to include the images in the layers of information on the DT.

Similar to the 360 images, during the recording of the videos people were in the scenes. Therefore, to comply with GDPR requirements, before using the videos on the DT, all the people appearing in the videos were blurred using out-of-the-shelf computer vision techniques. Finally, to be able to use the videos in the DT, they were hosted on the YouTube platform where a public link is then generated.

To create a mesh model of the newly renovated area in DCU Glasnevin, all the drone images were fed into the software iTwin Capture Modeler. This new model and the old one are both geolocated, which is desirable since this geolocation information is used to stitch both maps to create a new updated version of the DCU Glasnevin campus containing now the new high-tech Polaris building.

DCU Campus Explorer is to be a live system accessed by any person interested in DCU. As such, all data shown in this service must be authorized by DCU. For now, the only real-time sensor information presented in this system is for bus stops around the campuses. This system was developed by SimplyTransport[1] and this work is leveraging its API to enrich DCU Campus Explorer by displaying the dashboard of the bus stops that are close to DCU.

4.3 DCU Campus Explorer Data Integration and Accessibility

The current DCU DT has been integrated on two different platforms OCP (OpenCities Planner) and UE (Unreal Engine). Each platform offers different levels of accessibility and immersion. For example, OCP offers great accessibility as it can be accessed from the web and used on different devices such as phones, tablets and laptops. But with this accessibility it only offers a medium level of immersion. On the other hand, UE provides a high level of immersion such as in games, but it can only be accessed on high-end computers or hosted on expensive services. Each platform has its own advantages and disadvantages. However, looking back to the requirements from DCU C&C. They need a Campus Explorer that is easily accessible from the web and from mobile devices such as mobile phones, tablets and laptops. In that sense, OCP was selected. OCP offers a range of functionalities and for each of the requirements certain functionalities were leveraged as described below.

OCP offers a functionality called viewpoints that allows the setup of a list of important views of each campus, as presented in Fig. 4. This functionality was used to create a list of views to fulfill the requirement of having quick campus overview.

The second requirement for overlaying information about buildings was completed using points of interest, groups and layers on OCP. This information was organized in a group called Buildings and under three different layers called Campus Entrances, Building Entrances and Buildings, as presented in Fig. 5.

Information about services and facilities was put in a group called Points of Interest and it was organized under 20 layers such as Advice and Services, Be Active, Connect with Others, Fig. 6. Inside each layer, all information related to it was integrated as POIs (Points of Interest). In OCP each POI can be accompanied by a title, a description, images, and a starting view, as presented in Fig. 7. The description field of the POIs allows to write source code, therefore using this flexibility iframes (inline frame) were used to include the link to the 360 images which are hosted at momento360.

Besides POIs, OCP has the functionality of 3D Model which allows to introduce basic 3D forms such as cylinders and polygons. And the functionality of Terrain Texturing which allows to create lines on top of the 3D mesh map of the campus. To create the routes, these two functionalities were leveraged. A route is created using a cylinder showing the start of the route, a line showing the path of the route, and finally a polygon showing the end of the route, as presented in Fig. 8. To complete the routes with visual

---

[1] https://simplytransport.ie.

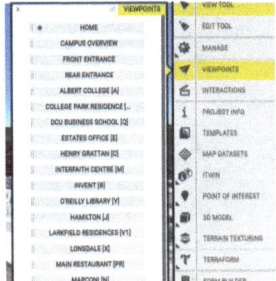

Fig. 4. Quick overview.

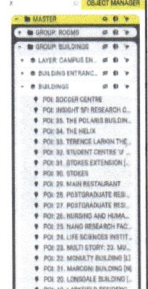

Fig. 5. Buildings.

Fig. 6. Services and facilities.

information videos were added using the source code functionality in the description of the layers. Like the 360 images, iframes were also used.

The last requirement was fulfilled using the same groups, layers, and POIs functionalities but in this case, they were put in a numeric order in which they have to be visited, as seen in the Fig. 9. For now, three types of tours were defined, for students, staff and history hounds.

Fig. 7. POIs functionality.

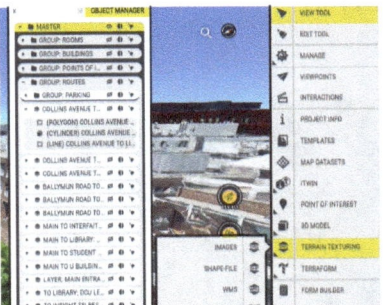

Fig. 8. Elements of routes.

Fig. 9. Tours.

5   Results and Discussion

This section presents the created Campus Explorer using the available technology and resources from Insight and DCU which overlays the information created by DCU C&C. This system is now live running to be accessed for 22,000 students, more than 3,000 staff plus visitors at https://www.dcu.ie/CampusExplorer.

To be friendly with the users and to maximize the usability of the Campus Explorer, a web page was created on the DCU website to present an explanation about it. Here the users find an overview of the design features, a demo on how to use the Campus Explorer, and finally a link to the three campuses, Fig. 10.

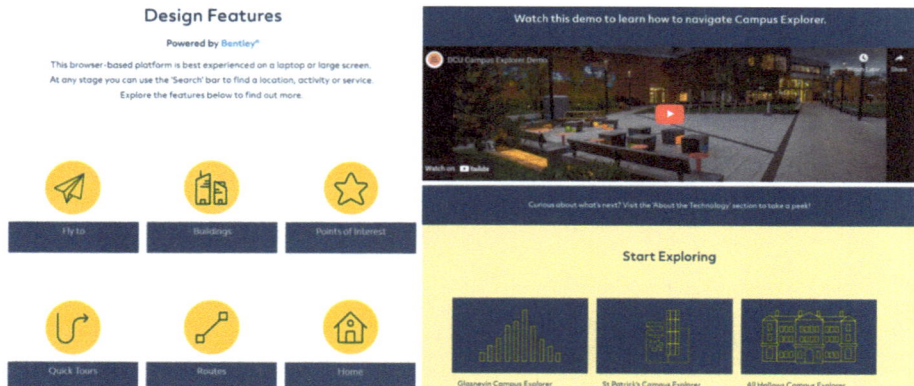

Fig. 10. DCU Campus Explorer web page.

The DCU Campus Explorer presents five main groups of information on the right side of the screen, each of them is related to each requirement set by DCU C&C.

To start, to have a campus overview, users can use the **Fly To** group; by clicking any of the views presented on it, the user will be taken to that specific view (Fig. 11). On a mobile phone this is presented on the top of the screen. The users can click the left/right arrow to go to the views or select a specific view from the dropdown menu by clicking Navigation, Fig. 12.

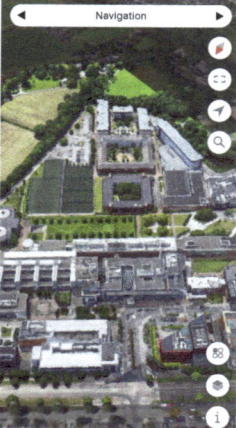

Fig. 11. DCU campus overview.

Fig. 12. Mobile view.

Information about buildings was set in the **Building** group. By clicking it, the users will be able to see the campus entrances, building entrances and the name of the buildings overlayed on top of the 3D campus map. Figure 13 presents the labels of the buildings on the Glasnevin Campus (left), the building entrances on the All Hallows (top right), and the entrances of the St. Patrick's campus (bottom right).

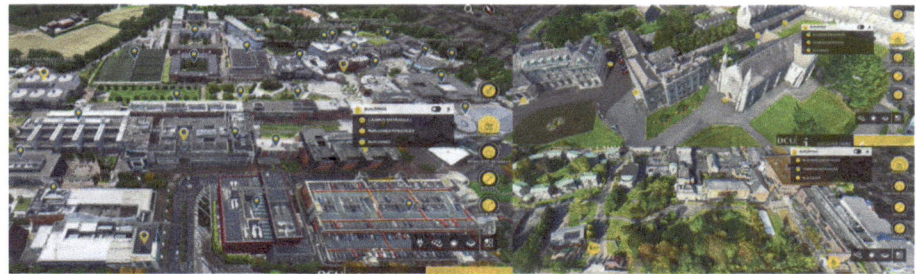

Fig. 13. Building information.

The information on services and facilities was set in a group called **Points of Interest**. Inside this group one layer per category was created. In this case when the user clicks a specific layer, all the POIs related to that category is displayed which then the user can select individually to see the relevant information, as presented in Fig. 14.

Information about routes can be accessed in the group **Routes.** As planned, the implementation of the routes, besides having an overlay of the path on top of the 3D models, videos were attached to them to have a better visualization of the routes. A clear example of this can be seen in Fig. 15.

Finally, to see the predefined tours on the campuses, the group **Quick Tours** is available. The users can click the places in numerical order and the interface will take them to each of them, Fig. 16.

Fig. 14. Services and facilities information.

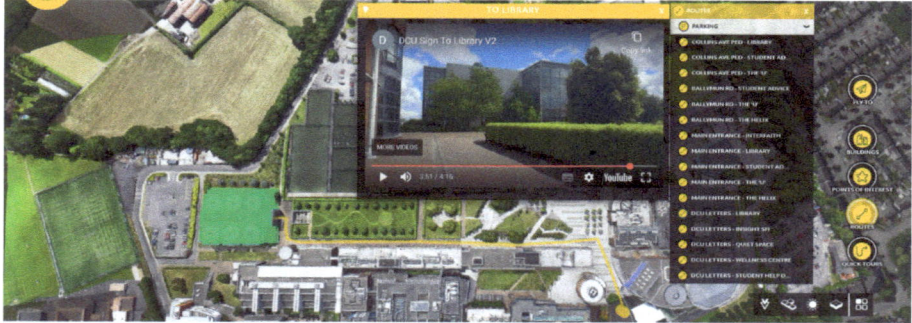

Fig. 15. Routes information.

Fig. 16. Quick tours information.

6 Conclusion

This work presents the application of DTs to fulfill a real-life case scenario. An application that resulted in the creation of a live Campus Explorer that the DCU community can use to access DCU information of services and facilities in a friendly and organized manner.

The mentioned methodology, besides integrating 3D modeling, it also focuses on the collection and the creation of layers of contextual information. Information that is then overlayed on top of the produced 3D models as POIs. All this with the intention of providing spatial context. OCP was used to integrate the different sources of information since it provides great accessibility and friendly interfaces. This methodology shows that current DT technology can provide a complete solution for citizen engagement in university environments by having the flexibility to integrate different sources of information and make it accessible in a user-friendly manner.

Besides bus stop timetables, no real-time IoT sensor data is displayed in the current Campus Explorer. However, it can be easily integrated in the future once proper agreements are reached by DCU. The other limitation is that this Campus Explorer has only the role of disseminating the desired information from DCU C&C to the DCU community but not functionality to get feedback from the DCU community to DCU C&C is implemented yet.

Regarding the mentioned limitations, future work will focus on the development of an interface to collect feedback from the DCU community to improve current services and facilities of DCU. It is also planned to create a structured database with the collect information that can be accessed from any integration tool besides OCP.

Acknowledgments. This publication has emanated from research conducted with the financial support of Taighde Éireann – Research Ireland [12/RC/2289_P2] at Insight Research Ireland Centre for Data Analytics, Dublin City University; The authors also thank Bentley Systems for their support and funding.

Disclosure of Interests. The authors have no competing interests to declare that are relevant to the content of this article.

References

1. Citizen Engagement: https://www.publicconsultation.ie/citizen-engagement. Last accessed 1 Mar 2025
2. Civic and Citizen Engagement: https://www.worldbank.org/en/topic/citizen-engagement. Last accessed 25 Mar 2025
3. Huttunen, S., Ojanen, M., Ott, A., Saarikoski, H.: What about citizens? A literature review of citizen engagement in sustainability transitions research. Energy Res. Soc. Sci. **91**, 102714 (2022)
4. Daher, E., Maktabifard, M., Kubicki, S., Decorme, R., Pak, B., Desmaris, R.: Tools for citizen engagement in urban planning. Holistic Approach for Decision Making Towards Designing Smart Cities, pp. 115–145 (2021)
5. Gonzalez-Mohino, M., Rodriguez-Domenech, M.Á., Callejas-Albiñana, A.I., Castillo-Canalejo, A.: Empowering critical thinking: the role of digital tools in citizen participation. J. New Approaches Educ. Res. **12**(2), 258–275 (2023)
6. Shin, B., et al.: A systematic analysis of digital tools for citizen participation. Gov. Inf. Q. **41**(3), 101954 (2024)
7. Berigüete, F.E., Santos, J.S., Rodriguez Cantalapiedra, I.: Digital revolution: emerging technologies for enhancing citizen engagement in urban and environmental management. Land **13**(11), 1921 (2024)
8. Kaunas digital twin created by KTU is among the best in the world. Kaunas University of Technology. https://en.ktu.edu/news/kaunas-digital-twin-created-by-ktu-is-among-the-best-in-the-world/. Last accessed 3 Feb 2025
9. Opoku, D.G.J., Perera, S., Osei-Kyei, R., Rashidi, M., Bamdad, K., Famakinwa, T.: Digital twin for indoor condition monitoring in living labs: University library case study. Autom. Constr. **157**, 105188 (2024)
10. RealSim Virtual Campus. https://www.universityofgalway.ie/about-us/press/publications/e-scealaapril2013/realsimvirtualcampus/. Last accessed 20 Feb 2025
11. U. of Birmingham, University of Birmingham. Virtual Tour, https://www.birmingham.ac.uk/virtual-tour. Last accessed 21 Feb 2025
12. Han, X., et al.: Intelligent campus system design based on digital twin. Electronics **11**(21), 3437 (2022)
13. Ye, X., Jamonnak, S., Van Zandt, S., Newman, G., Suermann, P.: Developing campus digital twin using interactive visual analytics approach. Front. Urban Rural Plann. **2**(1), 9 (2024)

14. García-Aranda, C., Martínez-Cuevas, S., Torres, Y., Pedrote Sanz, M.: A digital twin of a university campus from an urban sustainability approach: case study in Madrid (Spain). Urban Sci. **8**(4), 167 (2024)

15. Pavón, R.M., Alberti, M.G., Álvarez, A.A.A., Cepa, J.J.: Bim-based Digital Twin development for university Campus management. Case study ETSICCP. Expert Syst. Appl. **262**, 125696 (2025)

16. Fernandez, J.B., et al.: Smart DCU digital twin: towards smarter universities. In: 2023 IEEE Smart World Congress (SWC), pp. 1–6. IEEE (2023)

17. Fernandez, J.B., Ali, M.I.: System demo of modeling smart university campus virtual environments. In: International Conference on Multimedia Modeling, pp. 218–224. Springer Nature Singapore, Singapore (2025)

Author Index

© The Editor(s) (if applicable) and The Author(s) 2026
A. Jurelionis et al. (Eds.): BDTIC 2025, LNCE 775, pp. 179–180, 2026.
https://doi.org/10.1007/978-3-032-09040-9

Zeitfracht Medien GmbH
Ferdinand-Jühlke-Straße 7
99095 Erfurt, Deutschland
produktsicherheit@kolibri360.de